Djamila BAAZIZE-AMMI
Nadia HEZIL

FAT AND PROTEIN CONTENT OF MILK (Processing and quality)

Djamila BAAZIZE-AMMI
Nadia HEZIL

FAT AND PROTEIN CONTENT OF MILK (Processing and quality)

Milk component technology

ScienciaScripts

Imprint
Any brand names and product names mentioned in this book are subject to trademark, brand or patent protection and are trademarks or registered trademarks of their respective holders. The use of brand names, product names, common names, trade names, product descriptions etc. even without a particular marking in this work is in no way to be construed to mean that such names may be regarded as unrestricted in respect of trademark and brand protection legislation and could thus be used by anyone.

Cover image: www.ingimage.com

This book is a translation from the original published under ISBN 978-620-6-72045-4.

Publisher:
Sciencia Scripts
is a trademark of
Dodo Books Indian Ocean Ltd. and OmniScriptum S.R.L publishing group

120 High Road, East Finchley, London, N2 9ED, United Kingdom
Str. Armeneasca 28/1, office 1, Chisinau MD-2012, Republic of Moldova, Europe
Printed at: see last page
ISBN: 978-620-8-13123-4

Preamble

In the vast realm of nutrition and gastronomy, milk has a place of honour. At the heart of this white, milky liquid are two essential ingredients: fat and protein. Although these two components are essential, their complexity and importance are often overlooked.

It is with this in mind that this book, entitled "Milk Fat and Protein: Transformation and Quality", has been published. Through its pages, we dive into the mysteries of the transformation of these fundamental elements of milk, exploring their journey from dairy farms to our plates.

Fat, the source of rich flavours and creamy textures, and protein, which provides the essential building blocks for our bodies, are at the heart of many culinary creations and industrial processes. A thorough understanding of them is therefore essential, whether for producers, processors, researchers or simple lovers of good food.

This book is the result of meticulous research and an in-depth summary, with the aim of making these two elements of milk easier to understand, while highlighting the challenges and advances that define their transformation process. If you're curious about the wonders hidden in every drop of milk, this book invites you on a captivating journey to the heart of milk fat and protein, their transformations and the never-ending quest for quality.

TABLE OF CONTENTS

INTRODUCTION

Milk, the nourishing liquid that has symbolised fertility and vitality for thousands of years, is much more than just a drink. Beyond its apparent simplicity, it is an essential source of nutrients for many populations around the world, providing a unique combination of macronutrients, such as proteins and fats, which are essential for a balanced diet. Two key components of milk are fat and protein, which play vital roles both nutritionally and functionally. Milk fat, mainly made up of triglycerides, gives milk its creamy texture and characteristic flavour, while providing a concentrated source of energy. Milk proteins, in particular casein and whey, are crucial for growth and development, as well as for maintaining muscle and immune health.

Processing these milk components is a complex process, involving a series of carefully orchestrated stages to obtain a variety of finished products tailored to consumers' needs. From traditional methods to modern technologies, dairy producers use a range of techniques to separate, purify and process milk fat and protein, while preserving their nutritional and functional properties as much as possible.

In this exploration, we'll delve into the world of milk fat and protein, examining their characteristics, functions and the challenges associated with processing them. From farm to table, we'll discover the innovations and practices that are shaping the modern dairy industry and influencing our daily experience with these essential elements of our diet.

In Algeria, a wide variety of dairy products are prepared using naturally fermented milk. These products still play a major role in people's diets, particularly in rural areas. Among them, Lben, Raib, Zebda, Dhan or Smen and Jben are the most widespread and are mostly marketed through informal channels.

Together, we will discover this captivating universe where science and tradition meet, where culinary passion meets technological innovation. Whether you are a dairy industry professional, an enthusiast, or simply curious to learn more about the mysteries of milk, this book offers you an immersion into the heart of milk's fat and protein components, their transformation and the ongoing quest for quality.

MAMMARY GLAND

1. Anatomy of the cow's udder

The mammary gland (GM) is a secretory organ specific to mammals. The function of lactation or milk production is linked to a long process of development and differentiation of mammary tissues. The mammary gland develops progressively from foetal life onwards, following the female's gestation and lactation cycles. The anatomy of the mammary gland differs slightly from one species to another. In cattle, it is made up of four anatomically and physiologically independent quarters. Like all farmed ruminants, these quarters have a cistern for storing milk between feedings or milkings (Figure 1).

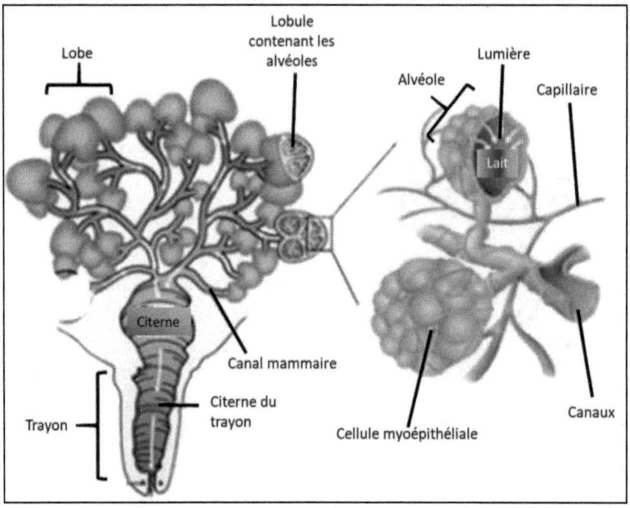

Figure 1: Diagram of the cow mammary gland and the structure of a mammary alveolus (Husveth, 2011).

Two types of tissue are present in the GM: secretory tissue or mammary alveoli and non-secretory tissue.

- Secretory tissue: The mammary epithelial cells (MECs) are joined together in a monolayer epithelium which forms the spherical structure of the alveoli (or acini) and ensures their secretory function (Figure 2). These cells rest on a basal lamina and are polarised. They take nutrients from the blood at their basal pole. They secrete the various milk constituents at their apical pole.

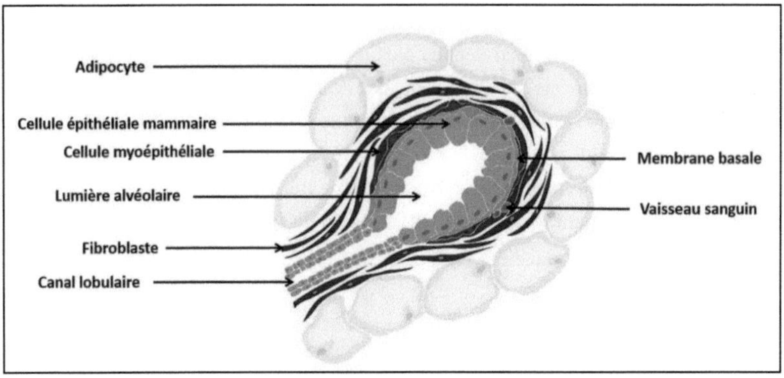

Figure 2: Structure of the mammary alveolus (Delouis and Richard, 1991).

- Non-secretory tissue or stroma: this is made up of connective tissue, adipose tissue, muscle, blood and lymph vessels and nerves.

o Connective tissue: plays a role in maintaining and supporting the mammary gland. which weighs between 14 and 32 kg when empty, and can exceed 50-60 kg in high-producing cows. It therefore requires solid support structures. There are two types of connective tissue in the udder: loose connective tissue is made up of fat cells surrounded by a matrix of fibres and fundamental substance, and dense connective tissue is mainly made up of ligaments and tendons.

o Muscular tissue. The alveoli and milk ducts are surrounded by smooth muscular tissue known as myoepithelial tissue. The contraction of myoepithelial cells under the effect of oxytocin is responsible for the ejection of milk.

o Blood and lymphatic vessels: The udder is irrigated by the external pudental artery and the perineal artery, which are subdivided into an increasingly fine network. The venous collection system is made up of the external pudendal vein, the cranial mammary vein and the internal pudendal vein.

The udder also has a lymphatic system that allows a large influx of neutrophils. The lymph passes through the retro-mammary lymph nodes.

o Nerves: The udder is innervated by the first 4 pairs of lumbar nerves. The major udder nerve is the genitofemoral nerve.

o Skin: The skin plays a minor role in supporting and stabilising the udder, and is held to the connective tissue by elastic fibres.

Teat: this is the junction between the glandular cistern and the papillary cistern and consists of an annular relief and a proximal venous plexus called the Fürstenberg venous circle. The papillary cistern communicates with the outside via a short papillary duct (0.8-1.0 cm) opening into the papillary orifice. The papillary duct is longitudinally folded and has a ring of folds at its proximal end called the Fürstenberg rosette. It is covered by a multi-layered epithelium. The papillary orifice is surrounded by muscle fibres forming the papillary sphincter, which is important for the regularity of the milk flow (Figure 3).

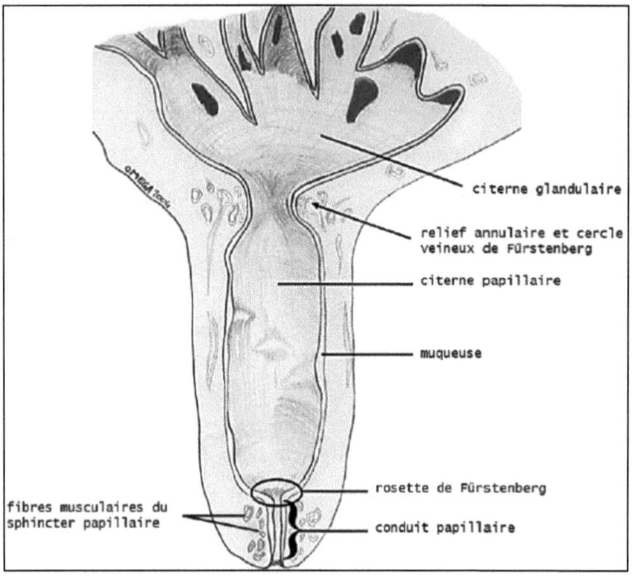

Figure 3: Anatomy of the teat (Couture and Mulon, 2005)

2. Mammogenesis

Milk production is the result of a long process of development and differentiation of mammary tissue. From embryonic life to the first lactation, the mammary gland goes through a series of hormone-controlled stages of development known as mammogenesis (development of the glandular parenchyma, i.e. cell multiplication and establishment of the lobuloacinar organisation).

During lactation there are three phases: lactogenesis, galactopoiesis and involution.

Lactogenesis and galactopoiesis (two phases linked to milk production) are also controlled by hormones: prolactin, glucocorticoids and growth hormone. The proliferative and regenerative capacities of mammary tissue during reproductive cycles and lactations make the mammary gland a unique organ, enabling it to renew 50% of its cells during a lactation.

- **Lactogenesis** is a phase of cellular differentiation and the acquisition of milk synthesis and secretion activities which takes place just before parturition. From day 150ème of gestation, the lobulo-alveolar system is established, gradually replacing adipose tissue. Ten days before parturition, the mammary epithelial cells hypertrophy and acquire the structures specific to synthesis and secretion: a significant increase in the volume of the cytoplasm, migration of the nucleus to a basal position, a highly developed rough endoplasmic reticulum, a Golgi apparatus associated with numerous vesicles destined for secretion, and a large number of mitochondria (Figure 4). The quantity and quality of parenchymal tissue are determining factors in milk production capacity.

- **Galactopoiesis** is the production of milk by the mammary glands, leading to lactation. Milk is produced and secreted continuously. This process is self-sustaining but declines over time. At its peak, the epithelium can synthesise and secrete 15% of its own weight per day in the form of proteins. In cows, milk secretion increases rapidly from parturition to the peak of lactation, then gradually declines over time. Milk secretion is dependent on the number and activity of mammary epithelial cells (MECs). After peak lactation, the loss of milk production is due in part to an imbalance between the proliferation and progressive elimination of MECs. The latter is induced by two factors: the entry of secretory cells into apoptosis and their exfoliation.

- **Involution:** corresponds to the resting of the udder and therefore to the dry period of lactation.

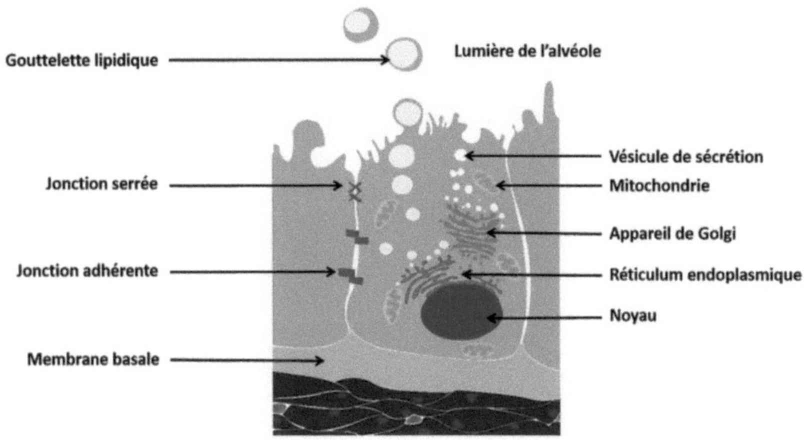

Figure 4: Organisation of a differentiated epithelial cell (Jammes and Djiane, 1988)

MILK

1. Definition

Milk is a liquid twice as viscous as water, with a slightly sweet taste and a mild odour. Its taste varies according to animal species, and is pleasant and sweet.

According to the International Congress of Fraud Control, it is the entire product of the total and uninterrupted milking of a healthy, well-fed and not overworked dairy cow. It must be collected cleanly and must not contain colostrum.

According to the Codex Alimentarius, milk is the normal mammary secretion of milking animals obtained from one or more milkings, without anything added or subtracted, intended for consumption as liquid milk or for further processing.

2. Composition of milk

Milk is an extremely complex biological fluid. The composition of milk from different species of placental mammals contains a number of common components, including caseins aggregated into micelles, triglycerides secreted in the form of fat globules, lactose and calcium. However, the fine composition varies quantitatively and qualitatively from species to species, and is influenced by genetic and husbandry factors. These differences in composition largely determine the nutritional, sensory and technological qualities of milk.

Ruminant milks have a high protein content and are also characterised by a high proportion of short-chain fatty acid (FA) residues in the triglycerides. Cow's and goat's milk have the most balanced lipid, lactose and protein compositions (Table 1).

Table 1: Average composition of milk from different animal species (Amiot et *al*., 2002)

Animals	Water(%)	Fat content(%)	Protein (%)	Carbohydrates (%)	Minerals (%)
Cow	87,5	3,7	3,2	4,6	0,8
Goat	87,0	3,8	2,9	4,4	0,9
Ewes	81,5	7,4	5,3	4,8	1,0
Chamelle	87,6	5,4	3,0	3,3	0,7
Mare	88,9	1,9	2,5	6,2	0,5

2.1. Water

Milk contains an average of 875g/l of water, the most important element. It determines the different phases of the milk

- Real solution containing sugars, soluble proteins, minerals and water-soluble vitamins
- Colloidal solution containing proteins, in particular caseins
- Emulsion of fats in water

This water can be found in two states:

- Extra micellar water: represents 90% of total water, and contains almost all the lactose, soluble mineral salts and soluble nitrogen. A small proportion of this water is bound to water-soluble elements, including soluble proteins.

- Intra-micellar water: represents around 10% of total water. A fraction of this water is bound to the caseins and the rest retains its solvent properties.

2.2 Carbohydrates

Lactose is the main carbohydrate in milk. It partly regulates the volume of milk produced thanks to its osmotic power. Lactose, a disaccharide made up of glucose and galactose, is the only free carbohydrate in milk present in significant quantities, with a very stable content of between 48 and 50 g/L. Unlike the butyrous content, lactose content varies only slightly. Lactose plays a special nutritional role and is involved in fermentation. It can be hydrolysed by strong acids, but especially by lactase. Cow's milk also contains around forty oligosaccharides. These are increasingly being studied for their role in the nutritional quality of milk.

2.3 Minerals

Minerals (or salt matter) are present in milk at around 7g/litre (Table 2). The most abundant are calcium, phosphorus, potassium and chlorine. These saline substances are found either in solution in the soluble fraction, or in bound form in the insoluble (or colloidal) fraction. Some minerals are found exclusively in a dissolved state in the form of ions (sodium, potassium and chlorine) and are particularly bioavailable. The others (calcium, phosphorus, magnesium and sulphur) exist in both fractions. In the soluble fraction, they exist partly in free form (ionised calcium and magnesium), partly in saline form (phosphates and citrates) undissociated (calcium and magnesium), or in complex form (phosphoric esters and phospholipids). In the colloidal fraction, the minerals (calcium, phosphorus, sulphur and magnesium) are associated or bound to the casein within micelles. Milk is a poor source of trace elements. They are usually found at relatively low levels.

Table 2: Mineral composition of milk (Amiot et *al.*, 2002).

Minerals	Content (mg/kg)	Trace elements	Content (mg/kg)
Calcium (Ca)	1180	Zinc (Zn)	3,80
Potassium (K)	1500	Iron (Fe)	0,50
Chlorine (Cl)	958	Iodine (I)	0,28
Phosphorus (P)	896	Copper (Cu)	0,10
Sodium (Na)	445		
Magnesium (Mg)	105		

Added to this list are certain elements, such as the sulphur present in proteins and the following trace elements, which are present in low concentrations in trace amounts: manganese, boron, fluorine, silicon, bromine, molybdenum, cobalt, barium, titanium, lithium and probably some others.

2.4. Vitamins

On the one hand, there are the water-soluble vitamins (group B vitamins and vitamin C) in constant quantities. Vitamin B, and more specifically B12, is more abundant in ruminants than in monogastric animals because of its microbial origin. The other vitamins are fat-soluble (A, D, E and K) (Table 3). As these vitamins are dissolved in fat, they are passed on during the skimming process in cream and butter, and are not very present in skimmed-milk products.

Table 3: Vitamin concentrations in cow's milk (FAO, 1998)

Water-soluble vitamins	(mg/L)	Fat-soluble vitamins	(mg/L)
B1 (thiamine)	0.42	A	0,37
B2 (riboflavin)	1 ,72	ß-carotene	0,21
B6 (pyridoxine)	0,48	D (cholecalciferol)	0,0008
B12 (cobalamin)	0,0045	E (tocopherol)	1, 1
Nicotinic acid (niacin)	0,92	K	0,03
Folic acid	0,053		
Pantothenic acid	3,6		
Biotin	0,036		
C (ascorbic acid)	8		

2.5. Proteins

The protein content corresponds to the total nitrogen content of the milk. Milk proteins account for 95% of nitrogenous matter. The remaining 5% is made up of free AA, small peptides, non-protein nitrogen (mainly urea), creatinine and uric acid. There is little variation in the protein content of milk, ranging from approximately 30 to 35 g/kg.

There are two main categories of milk protein: spherical casein micelles, which are non-soluble and measure between 30 and 600 nm (180 nm on average), and serum proteins, which are soluble. The different proteins are shown in Table 4.

Table 4: Protein composition of cow's milk (Jensen, 1995).

Proteins	Concentration (g/L)
Caseins	
αs1	10.0
αs2	2.6
β	9.3
κ	3.3
γ	0.8
Soluble proteins	
β-lactoglobulin	3.2
α-lactalbumin	1.2
BSA (Bovine Serum Albumin)	0.4
Immunoglobulins (G, A and M)	0.8
Proteose peptones 8 (fragments of β-casein)	0.5
Lactophorin	0.3
Lactoferrin	0.1

Transferrin	0.1
Membrane of fat globules	0.4

Caseins are phosphoproteins organised in milk in the form of micelles. There are around 1.14 x 10^7 casein micelles per litre of milk, with an average size of 150 to 200 nm.

In cattle, there are four types of casein: αS1, αS2, β and κ. Micelles also contain minerals, mainly calcium phosphate (90%) and magnesium and citrate ions (10%). Although the caseins inside the micelle are bound together by calcium phosphate and the κ-caseins are predominantly present on the surface (Figure 5), the exact structure of the casein micelle is not yet unanimously agreed.

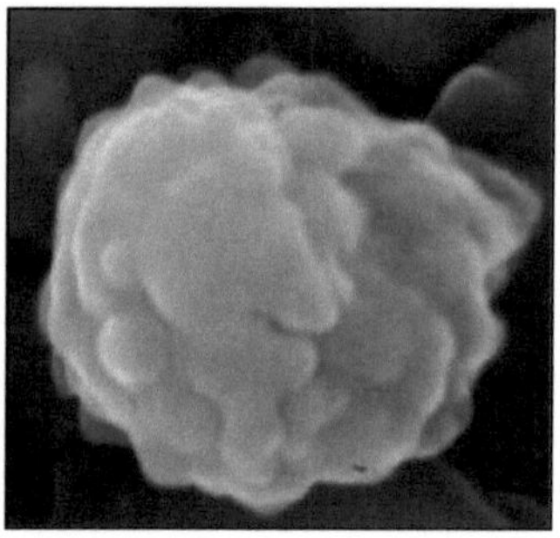

Figure 5: Structures of native milk casein micelles obtained by scanning electron microscopy at magnification x70000 (McMahon and McManus, 1998).

2.6. Fats and oils

In dairy cows, milk fat content can vary widely, from 30 to 60 g/L, but is usually between 35 and 47 g/L. The fat comes in the form of fat globules (FG) in emulsion in the aqueous phase of the milk. Cow's milk contains an average of 15 x 10^9 GGs and their diameter varies from 0.2 to around 15 μm (Figure 6). The size of the GGs is very important because it affects the technological and sensory properties of dairy products.

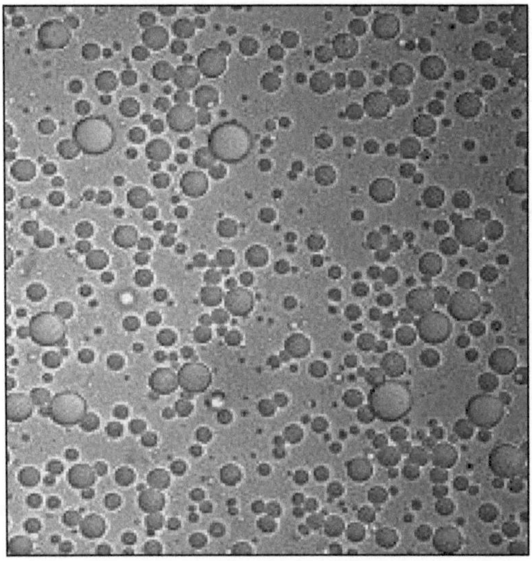

Figure 6: Fat globules in native milk under light microscopy (Lopez et *al.*, 2010).

The lipids in cow's milk are 98% triglycerides or triacylglycerols (TAGs), but milk also contains very small amounts of non-esterified free fatty acids, polar lipids (glycerophospholipids and sphingolipids), sterols, carotenoids and fat-soluble vitamins (Table 5). Despite their low quantity, these lipid compounds play an important functional role in native milk, stabilising fat globules by forming a membrane at the interface.

Table 5: Average composition of milk fat and composition of the main fatty acids making up TAGs (Croguennec et *al.*, 2008; Jensen, 2002)

Compound	**Content (%)**	**Main location**
Triacylglycerols (TAG)	97,5	*inside the fat globules*
Diacylglycerols	0,36	*inside the fat globules*
Monoacylglycerols	0,027	*at the interface (native membrane)*
Free fatty acids	0,027	*at the interface (native membrane)*
Phospholipids	0,6	*at the interface (native membrane)*
Cholesterol	0,31	*at the interface (native membrane)*
Fat-soluble vitamins	0,01	-
Carotenoids	0,008	-
Hydrocarbons	Traces	-

Milk TAGs are formed from over 400 different fatty acids. Cow's milk contains 15 major fatty acids (Table 6). The fatty acid (FA) composition of milk is complex. These fatty acids are divided into three families according to the number of double bonds in their carbon chain. Saturated fatty acids have no double bonds, while monounsaturated and polyunsaturated fatty acids contain one and several respectively. The double bond can be in *cis* or *trans* configuration, depending on whether the two parts of the carbon chain around the double bond are on the same side (*cis*) or on either side of the plane of the double bond (*trans*). GAs largely determine the physical properties of milkfat (melting, crystallisation, texture).

Table 6: Average concentration of the 15 main fatty acids in cow's milk (Jensen, 2002; Chilliard, 2006).

Fatty acid	**Nomenclature**	**Concentration (g/100 g of total GA)**
Butyric	C4:0	2-5
Caproic	C6:0	1-5
Caprylic	C8:0	1-3
Caprique	C10:0	2-4
Laurique	C12:0	2-5
Myristic	C14:0	8-14
Pentadecanoic	C15:0	1-2
Palmitic	C16:0	22-35
Palmitoleic	C16:1 cis-9	1-3
Margarique	C17:0	0,5-1,5
Stearic	C18:0	9-14
Oleique	C18:1 cis-9	20-30
Vaccenics	C18:1 trans-11	1-3
Linoleum	C18:2 cis-9, cis-12	1-3
Linolenic	C18:3 cis-9, cis-12, cis-15	0,2-1,5
AG satures		68,5

AG monoinsatures	29
Polyinsaturated GA	2,9

2.7. Other milk components

- **Milk metabolites**

Metabolites are intermediate molecules from metabolic pathways such as glycolysis, the Krebs cycle, respiration, the pentose phosphate pathway, etc. More than 40 different metabolites have been identified in cow's milk using Nuclear Magnetic Resonance (NMR). Other methods such as fluorimetry have been used to quantify major milk metabolites such as glucose-6-phosphate, isocitrate, glutamine, glutamate, etc.

- **Deoxyribonucleic acid (DNA) and Ribonucleic acid (RNA)**

The DNA present in milk comes mainly from somatic cells, including exfoliated MECs and bacteria that can be found in milk. It is detected by Polymerase Chain Reaction (PCR) tests. There are also different types of RNA, such as messenger RNA (mRNA). These RNAs are found in different fractions of milk, such as GG.

- **Somatic cells**

Milk contains predominantly immune somatic cells, leukocytes, including neutrophil polyporphonuclear cells (PMN), lymphocytes and macrophages. In cows, MECs are a minority in milk, with their relative proportion of the cell population ranging from 2 to 45%. The presence of immune cells in the milk is an indicator of mammary inflammation (mastitis), generally of infectious origin.

- **Bacteria**

In healthy animals, milk contains few micro-organisms when sampled under good conditions (less than 5,000 germs/ml and less than 1 coliform/ml). These are essentially saprophytic germs of the udder and milk ducts. However, in sick animals, other micro-organisms can be found in the milk and are generally pathogenic.

Bacteria can be classified according to their behaviour and the effects they generate. Six groups can be distinguished: lactic flora, heat-resistant flora, coliform flora, psychrotrophic flora, butyric flora and pathogenic flora.

3. Physico-chemical characteristics of milk

The main physicochemical properties used in the dairy industry are density, freezing point, boiling point and acidity.

3.1 Density

The density of a liquid is defined as the quotient of the mass of a certain quantity of the liquid divided by its volume. As density is closely dependent on temperature, it is necessary to specify the temperature (T) at which it is determined. The average density of whole milk at 20°C is 1030 $Kg.m^{-3}$.

3.2. Density

This value corresponds to the ratio of the mass of a volume of milk at a given temperature to that of the same volume of water at the same temperature. The mass ratio of cow's milk is generally between 1.023 and 1.040. It varies in the same direction as the milk's dry matter content.

3.3. Freezing point

The freezing point of milk is slightly lower than that of pure water, since the presence of solubilised solids lowers the freezing point. Its average value is between -0.54 and -0.55°C, which is also the freezing temperature of blood serum. This physical property is measured to determine whether water has been added to the milk. Wetting raises the freezing point towards 0°C, since the number of molecules other than water and ions per litre decreases. Generally speaking, any treatment of the milk or change in its composition that results in a change in its quantity will lead to a change in its freezing point.

3.4. Boiling point

The boiling point is defined as the temperature reached when the vapour pressure of the substance or solution is equal to the applied pressure. As with the freezing point, the boiling point is influenced by the presence of solubilised solids. It is slightly higher than the boiling point of water, i.e. 100.5°C. This physical property decreases with pressure. This principle is applied in milk concentration processes.

3.5. Milk acidity

The acidity of milk is the result of natural acidity, due to casein, phosphate groups, carbon dioxide and organic acids, and developed acidity, due to lactic acid formed during lactic fermentation. The measurement of titratable acidity is commonly expressed in two ways, either as a percentage (%) of lactic acid equivalents, or in degrees Dornic (°D); 1°D represents 0.1 g/l of lactic acid. The acidity of milk should be between 14 and 18 °D. Fresh milk has an acidity of 18° D.

3.6 pH

The pH of cow's milk usually varies between 6.5 and 6.7 (at 20°C), so it is very slightly acidic. Unlike titratable acidity, pH does not measure the concentration of acidic compounds, but rather the concentration of H^+ ions in solution. The pH values represent the state of freshness of the milk, particularly in terms of its stability, since it is the pH that influences the solubility of the proteins, i.e. reaching the isoelectric point. Milk with a high developed acidity will have a pH lower than 6.6, because lactic acid is a strong enough acid to dissociate and lower the pH by a measurable value. Similarly, as colostrum is acidic, a milk pH that is too low can also be indicative of the presence of colostrum, i.e. milking too early after parturition. Conversely, in the case of mastitis, the pH increases because the milk contains basic substances. Two milks can therefore have identical pH levels, i.e. be in the same state of freshness, but have different titratable acidities.

SUMMARY OF THE MAIN CONSTITUENTS OF MILK

The constituents of milk have two possible origins, either in the blood (water, vitamins, minerals, serum albumin, immunoglobulins, etc.), or in the mammary gland via their synthesis by the MECs from nutrients in the blood. The synthesis of milk constituents is made possible by the coordinated adaptation of the metabolism of the various tissues and organs that supply GM with nutrients.

1. Secretory pathways

Constituents can be secreted via one of 5 secretion pathways in EMCs: (a) transmembrane, (b) exocytosis, (c) the lipid secretion pathway, (d) transcytosis and (e) paracellular.

a. In the membrane pathway, substances can cross the apical cell membrane (and for those derived directly from blood, the basolateral membrane). Examples: water, urea, glucose, Na, K and Cl.

b. In the Golgi pathway, secretory products are transported or sequestered by the Golgi apparatus and secreted into the milk space by exocytosis. Examples are casein, whey protein, lactose, citrate and calcium.

c. In the milk fat pathway, milk fat globules are released from the apex of the secretory cell surrounded by bimembrane (milk fat globule membrane); some cytoplasm sometimes included. Examples are milk fat, fat-soluble hormones and drugs, certain unknown growth factors (in the milk fat globule membrane) and leptin.

d. In transcytosis, vesicular transport involves various organelles, possibly, in some cases, also involving extrusion via the second pathway (b). Examples are immunoglobulins during colostrum formation, transferrin and prolactin.

e. In the paracellular route, there is a direct passage from the interstitial fluid to the milk.

In full lactation, pathways 1, 2, 3 and possibly 4 predominate

1.1 Filtered blood constituents

Directly filtered constituents pass from the blood into the lumen of the acini via the lactogenic cells of the mammary epithelium. This is the case for :

Water, urea, minerals and vitamins.

Certain proteins: serum albumin, transferrin, certain enzymes and anti-enzymes, and immunoglobulin G (IgG).

Amino acids, creatinine, uric acid, ammonia and creatine.

Some fatty acids are also transferred unchanged from blood to milk. These include 50% of fatty acids with 16 carbon atoms and fatty acids with more than 16 carbon atoms: they come directly from the diet and are transported by the blood in the form of non-esterified fatty acids, chylomicrons and very low density lipoproteins (VLDL).

1.2 Constituents synthesised by the udder

1.2.1. Lactose

Lactose is the main sugar in milk. It is a diholoside resulting from the combination of one molecule of glucose and one molecule of galactose. Glucose comes from the blood and galactose is synthesised from glucose in the udder.

Lactogenic cells have the ability to isomerise some of the glucose they take from the blood into galactose by linking the glucose molecule with a Uridine DiPhosphate (UDP) molecule, thus obtaining UDP-galactose. This reaction in the Golgi apparatus involves α-lactalbumin and UDP-galactosyl-transferase.

The osmotic power of lactose is high, causing water to enter the Golgi vesicles by osmosis in order to dilute the newly synthesised lactose. As a result, the lactose content of milk is relatively constant and the quantity of lactose synthesised is one of the main determinants of the quantity of milk exported by the mammary gland.

1.2.2. Proteins

The proteins present in milk come either from synthesis in the MECs or from plasma. Caseins αs1, αs2, β and κ, β-lactoglobulin and α-lactalbumin are synthesised *de novo* by the MECs in the polysomes of the rough endoplasmic reticulum from amino acids in the blood plasma. Once synthesised, the proteins travel between the endoplasmic reticulum and the Golgi apparatus. It is in the Golgi apparatus and the trans-golgi network that proteins undergo post-translational modifications (phosphorylation, glycosylation or sulphation). In particular, caseins are phosphorylated by casein kinases, enabling them to aggregate into micelles in the presence of calcium and inorganic phosphate. Proteins modified in this way are transported by secretory vesicles to the apical membrane of MECs and secreted by fusion with it. The main factor limiting their synthesis is the availability of AA from the bloodstream. These AAs are also a source of nitrogen and carbon and can therefore be used to synthesise other AAs or in certain metabolic pathways such as the Krebs cycle. γ-casein is not synthesised by EMCs. It results from the hydrolysis of the C-terminal part of β-casein by plasmin.

Plasma proteins (immunoglobulins, serum albumin) are secreted into the milk via transcytosis mechanisms, i.e. the proteins are endocytosed at the basolateral level of the MECs and then transported to the apical pole for secretion (Figure 7).

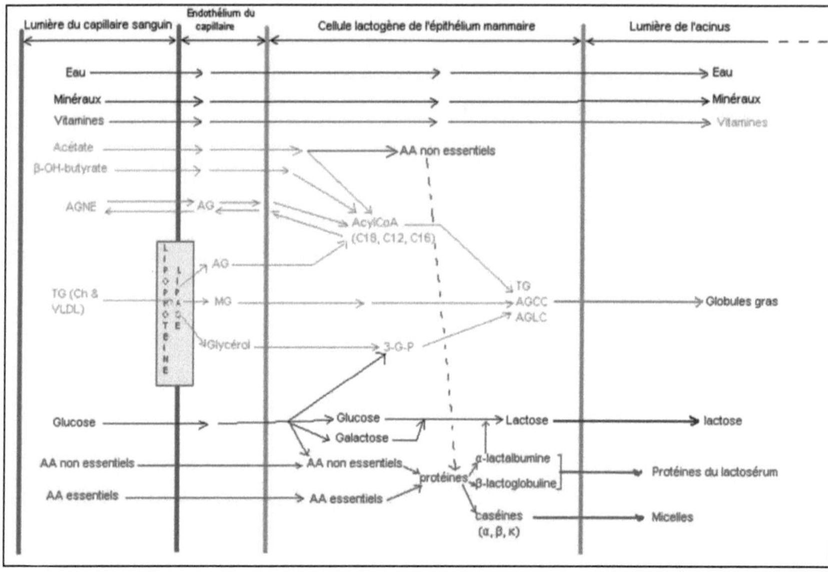

NEFA = non-esterified fatty acids; FA = fatty acids; AcylCoA (C18, C12, C16) = Acyl-coenzyme A; MG = monoglycerides; TG = triglycerides; Ch = chylomicron; VLDL = very low density lipoprotein; 3-G-P = 3-glycerol phosphate; AA = amino acids; SCFA = short-chain fatty acids; LCFA = long-chain fatty acids.

Figure 7: Global origin of milk constituents (Wattiaux, 1997)

1.2.3. Fat

Milk fat is particularly complex in terms of both composition and structure. There are no fewer than 7 complex families of lipids. In its native state, milk fat takes the form of fat globules (FG), which are droplets of triglycerides surrounded and stabilised in the aqueous phase of milk by a membrane.

Milk fatty acids come from both *de novo* synthesis in the mammary gland (40% of the fatty acids secreted in milk) and direct removal of plasma fatty acids and modification of part of the latter to give other fatty acids (60%) (Figure 8).

- ***De novo synthesis***

Lactogenic cells can synthesise certain fatty acids from ß-hydroxybutyrate and C2 (acetate) from rumen fermentation, the active forms of which in mammary epithelial cells are butyryl-CoA and acetyl-CoA respectively.

These two substrates are taken up by the enzyme acetyl-CoA carboxylase, which is responsible for activating acetyl-CoA to form malonyl-CoA. These GAs then undergo multiple condensation reactions by fatty acid synthase (FAS), which are interrupted by the enzyme thioesterase I to form short- and medium-chain saturated fatty acids (SFAs). *De novo* synthesis produces all the C6:0 to C12:0, 95% of the C14:0 and 50% of the C16:0 fatty acids in milk fat.

- ***Plasma sampling***

Plasma non-esterified fatty acids (NEFA) are extracted by the action of lipoprotein lipase (LPL) on the triglycerides present in intestinal chylomicrons and/or very low density lipoproteins (VLDL) from the liver. The NEFAs taken in also come from the mobilisation of body reserves (at the start of lactation). Of the milk FFAs, 50% of the C16:0 and all the FFAs with a carbon number greater than or equal to 18 come from plasma sampling.

- ***Desaturation and esterification of*** fatty ***acids***

Some of the medium-chain (C12:0, C14:0, C16:0) or long-chain (C18:0, *trans11-C18*:1, C20:0 to C24:0) FAs are desaturated in the Δ9 position by mammary Δ9-desaturase, the activity of which varies depending on the length of the FA under consideration and the isomer under consideration.

In the endoplasmic reticulum of the mammary secretory cells, the free GAs are esterified on a glycerol-3-phosphate molecule obtained directly from plasma or synthesised in the mammary gland from glucose. This step is carried out by three specific acyl-transferase enzymes, by successively adding a GA to one of the 3 positions of the glycerol-3-phospate to obtain a triglyceride.

The TGs are then transferred to lipid droplets before being secreted into the milk by exocytosis.

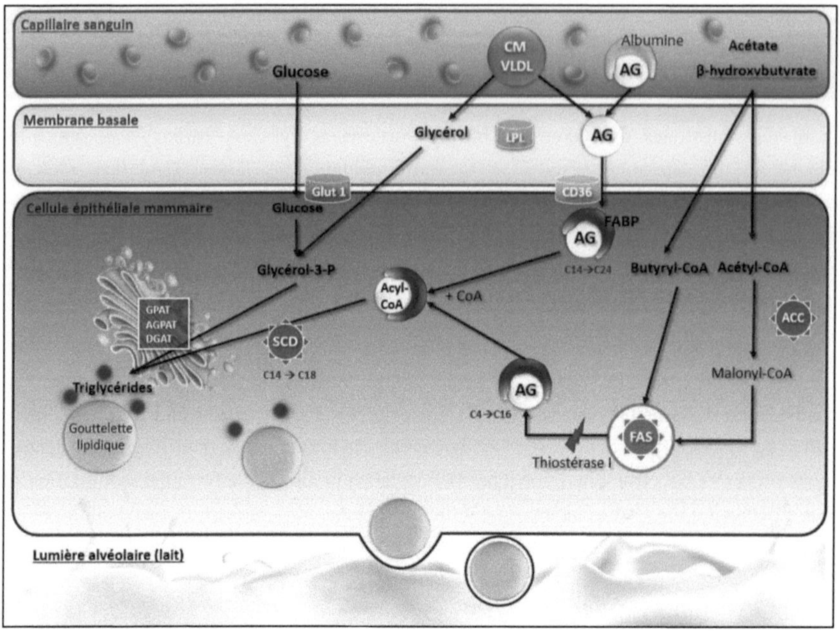

ACC, acetyl-CoA carboxylase; AGPAT, 1-acylglycerol 3-phosphate acyltransferase; CD36, differentiation group 36; CoA, coenzyme A; CM, chylomicron; DGAT, diacylglycerol acyltransferase 1; FA, fatty acid; FABP, fatty acid binding protein; FAS, fatty acid synthase; Glut 1, glucose transporter 1; GPAT, glycerol-3 phosphate acyltransferase; LPL, lipoprotein lipase; MFG, milk fat globule; SCD, stearoyl-CoA desaturase; TAG, triglyceride; VLDL, very low density lipoprotein.

Figure 8: Fatty acid metabolism in the mammary epithelial cell in ruminants: de novo synthesis, plasma uptake, esterification and secretion into milk (De Ofeu Aguiar Prado, 2018).

2. *Reminder of certain concepts*

o *Volatile fatty acids*: VFAs are fatty acids with a short carbon chain (less than six carbon atoms). The first four fatty acids are known as volatile: acetic acid (CH3-COOH), propionic acid (CH3-CH2-COOH), butyric acid (CH3-CH2-CH2-COOH), valerianic acid (CH3-CH2-CH2-CH2-CH2-COOH).

o *The different types of fatty acid according to their structure* (Figure 9).

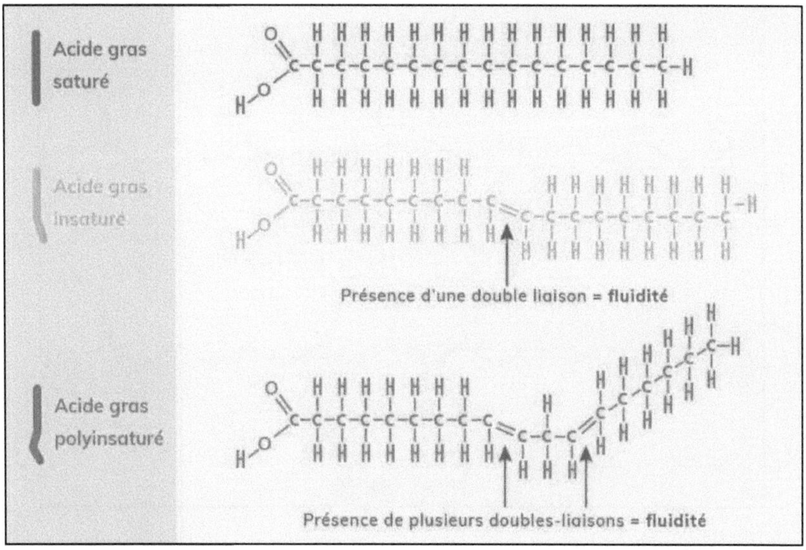

Figure 9: The different types of fatty acid according to their structures
https://www.pensersante.fr/acides-gras-satures-ou-insatures-quelle-difference

o *Trans and cis fatty acids*: Unsaturated fatty acids exist in either cis or *trans* form, depending on the position of the hydrogen atoms attached to the carbon atoms joined by double bonds.

If the hydrogen atoms are on the same side of the carbon chain, the arrangement is called *cis*.

If the hydrogen atoms are on either side of the carbon chain, the arrangement is called *trans* (Figure 10).

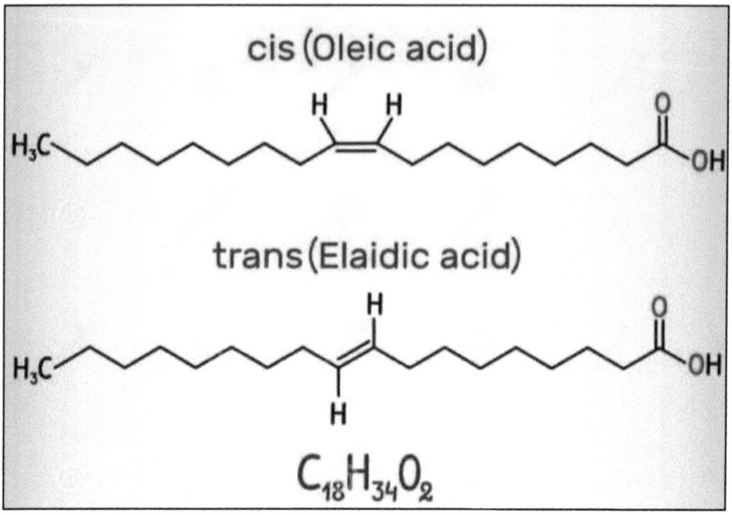

Figure 10: Trans and cis fatty acids.https://fr.dreamstime.com/transport-cis-%C3%A9la%C3%AFdique-acide-ol%C3%A9ique-d-les-acides-gras-omega-isom%C3%A8res-g%C3%A9om%C3%A9triques-formule-chimique-structurelle-image154963911

o *Fatty acid desaturase*: delta-desaturases are enzymes which act on a carbon-carbon bond whose position is defined in relation to the end of the hydroxyl group (OH) (e.g. stearyl-CoA9-desaturase or Δ9-desaturase enables the switch from saturated stearic acid to oleic acid, a monounsaturated fatty acid from the omega-9 family).

FAT PROCESSING

Milk fat is an essential component of dairy products, and its transformation is a process that produces a wide range of delicious and nutritious products.

1. Fatty Globule (GG)

1.1 Fat globule secretion

The synthesis and secretion of MG takes place within the mammary epithelial cell: the synthesis of MG takes place within the endoplasmic reticulum located in the basal region of the epithelial cell, then the MG transits towards the apex of the epithelial cell where it is secreted in the form of a GG. The triglycerides (TG) synthesised accumulate on the inner surface or within the reticulum membrane itself, forming microdroplets, so that when they are released into the cytoplasm, the droplets are covered by membrane (Figure 11). The growth and transit of the fat globule are two totally linked phenomena, since their growth takes place during transit from the basal to the apical region of the epithelial cell. On leaving the reticulum, the GG does not exceed 0.5 µm and reaches a size that varies to over 4 µm. The main phenomenon explaining this growth is the fusion of the microdroplets together. The GGs are secreted through a process of progressive envelopment by the apical plasma membrane. This process, known as apocrine secretion, is considered to be the majority.

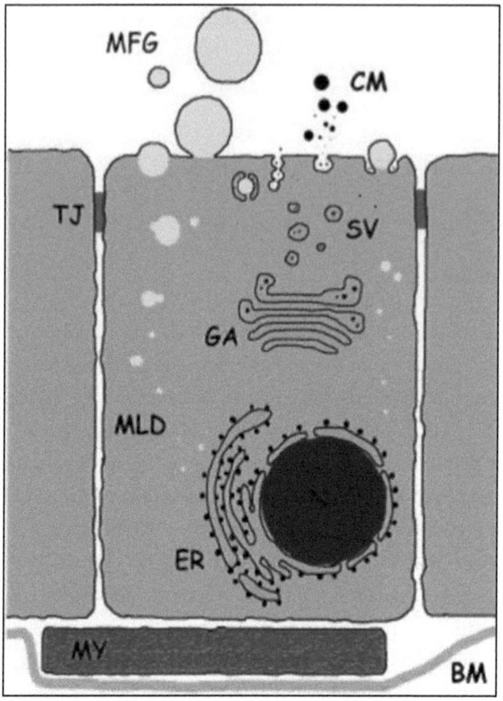

MFG: Fat globules; CM: Casein micelle; SV: Secretory vesicles; TJ: Tight junction; GA: Golgi apparatus; MLD: Lipid microdroplet; N: Nucleus; ER: Endoplasmic reticulum; MY: Myoepithelial cell; BM: Basal membrane.

Figure 11: Modified schematic representation of possible fat secretion pathways (Heid and Keenan 2005).

1.2 Structure of the fat cell

The GG membrane, which represents 2 to 6% of its total mass, is derived directly from cell membranes (smooth endoplasmic reticulum, golgi and apical membrane) and cytoplasmic contents. It is made up of three distinct parts

- The first, outermost part is a lipid bilayer derived directly from the apical membrane. Like other biological membranes, this bilayer has an asymmetrical phospholipid structure, with phosphatidylcholine and sphingomyelin mainly on the outer side of the bilayer and phosphatidylethanolamine and phosphatidylinositol on the inner side.

- The second, more internal part is thought to derive from the cytosolic fraction trapped between the lipid droplet and the apical membrane at the time of secretion. This part of the

GG membrane contains xanthine oxidase (12%) and butyrophilin (40%), the two main proteins in the GG membrane.

- The third part corresponds to a monolayer of proteins and polar lipids originating from the endoplasmic reticulum and already present on the surface of lipid droplets within the cell (Figures 12 and 13).

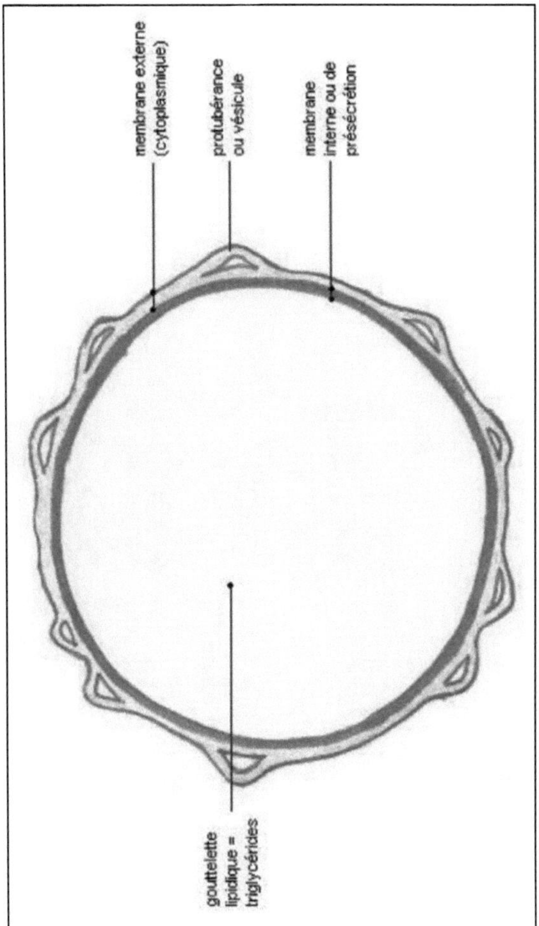

Figure 12: Fat globule (Mathieu, 1997)

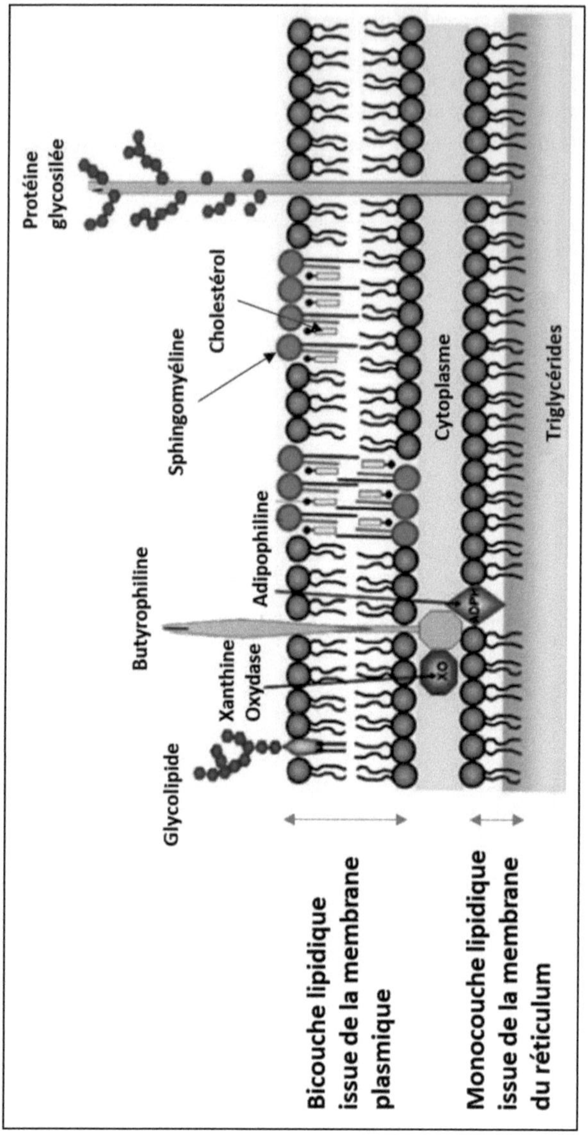

Figure 13: Structure of the fat globule membrane (Lopez, 2011).

1.3 Impact of processes on the composition of the fat globule membrane (MFGM)

1.3.1 Impact of heat treatment of milk

- ***Milk refrigeration***

During cold storage (4°C), the milk fat globule membrane undergoes certain changes. Some proteins (e.g. lactoferrin) associate more with the membrane, while others (e.g. β-casein) partially dissociate. As far as phospholipids are concerned, refrigeration can lead to the dissociation of a variable proportion of phospholipids. These losses are irreversible.

- ***Heating the milk***

Heating whole milk results in the loss of some MFGM proteins. A heat treatment lasting 10 minutes at 50°C results in the loss of around 50% of the proteins.

Heating whole milk also leads to losses of lipid components. Heat treatment for 20 minutes at 80°C would result in a significant loss of triacylglycerols (TAGs) and a negligible loss of phospholipids. However, several studies show that heat treatment of milk or cream results in the association of whey proteins with MFGM, mainly β-lactoglubulin.

1.3.2 Impact of mechanical treatments

- ***Churning***

Churning is a mechanical treatment which destroys the fat globules and releases their membrane material into the buttermilk.

Studies have shown that variables such as the pH of the cream or its temperature during churning can influence the proportions of non-membrane proteins found in extracts of MFGM fragments. Churning involves the incorporation of air for foam formation and a proportion of liquid fat must be present to allow phase inversion.

The incorporation of air into the milk causes significant damage to the MFGM. The fat globules attach themselves to the surface of the air bubbles and release their membrane material and part of their contents. When the air bubble is destroyed, the membrane material is released into the aqueous phase.

- ***Homogenisation***

The main effect of homogenising milk is to reduce the size of its fat globules. This results in an increase in the total surface area of fat. Since the membrane components are not present in sufficient quantity, milk proteins, mainly caseins, are used to stabilise the surface of the globules.

2. Transformation

2.1. Cream

Cream can be defined as an oil-in-water emulsion of dairy origin. Cream could be defined as milk 10 times more concentrated in fat. The term "cream" is reserved for milk containing at least 30% fat, exclusively of dairy origin, whether pasteurised or not.

2.1.1. Skimming

- ***Natural skimming***

Spontaneous skimming is a natural phenomenon whereby the cream rises slowly to the surface of the milk due to its difference in density to skimmed milk. This type of skimming is used in small artisan cheese dairies. It involves leaving the milk in a wide container, in a thin layer (10 to 15 cm) at a temperature of 8 to 14°C without stirring. After resting for 12 to 24 hours, the cream that has risen to the surface is collected using a flat ladle. This method of skimming is imperfect; under the best conditions, the quantity of fat in the cream does not exceed 80 to 85% of the milk fat. This creaming phenomenon is also due to the presence of agglutinins in the milk which, under the right conditions, cause the fat to clump together.

- ***Mechanical skimming***

The whole milk is first preheated to a temperature of up to 63°C and skimmed to obtain a cream with a fat content of around 30-55%. The centrifugal force of the skimmer speeds up the separation of the milk components: the denser components move to the outer walls while the lighter ones, the fats, gather in the centre to form the cream (Figure 14).

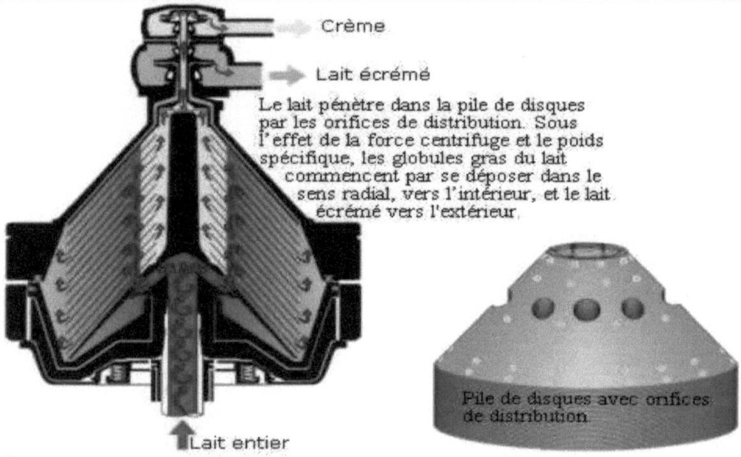

a

a: Diagram of a centrifugal skimmer bowl (vertical section), b: general view of a skimmer.

Figure 14: Centrifuge for skimming milk (Boutonnier, 2007).

- ***Factors influencing skimming efficiency***

o Particles suspended in the milk interfere with the separation of the fat globules. This is why, on arrival at the factory, the milk undergoes coarse filtration.

o The optimum skimming temperature is between 50 and 55°C. Raising the temperature has the dual advantage of reducing the viscosity of the liquid and increasing the difference between the densities of the fat phase and the skimmed milk.

o Air, present in the milk and incorporated during the various transfers, reduces the separating efficiency of the centrifuge and initiates and accelerates the fat alteration processes.

2.1.2. Pasteurisation and degassing of the cream

The cream is pasteurised using heat treatment scales of around 85-110°C for 10 to 30 seconds. This destroys pathogenic germs and most saprophytic germs, destroys lipases that cause rancidity, forms reducing sulphur compounds that prevent oxidation of the lipids, and subsequently controls the lactic ripening of the cream. Before or after pasteurisation, the cream may be deaerated. This air causes a loss of precision in volumetric measurements, as well as the risk of oxidation of unsaturated fatty acids. Cream may also contain malodorous substances: Originating from feed (wild plants in pasture, forage cabbage, etc); originating from the fixation, by the milk fat, of odours from various substances (hygiene products, various solvents, etc) and resulting from enzymatic or microbial activity. To do this, the cream is heated to a temperature of 78°C before being subjected to a vacuum which allows the evaporation of undesirable flavours and odours in a cyclone in which the cream circulates in a thin layer on the wall.

2.1.3. Biological maturation

To increase the viscosity of the cream and obtain a thick cream, it undergoes biological maturation. This maturation lasts between 15 and 20 hours. It takes place at low temperatures of around 14-15°C to encourage the aromatic microbial strains (*lactococcus* diacetilactis, *Lactococcus acetoïnus*, *Leuconostoc cremoris* and *Leuconostoc citrovorum*) which ferment the citrates and produce diacetyl, characteristic of butter flavour (hazelnut flavour), or higher towards 20-23°C in order to favour acidifying microbial strains (*Lactococcus lactis* and *lactococcus cremoris*) which transform lactose into lactic acid. The pH values vary according to the type of cream, ranging from 6.2 to 6.3 for fresh creams and 4.5 to 4.6 for acid creams. It is above pH 5.0 that the viscosity of the cream increases and the *Leuconostoc* bacteria develop, producing flavour. Fermentation lasts as long as is necessary to obtain the desired acidity (around 15 to 18 hours). A longer fermentation time enhances the flavour of the butter.

2.1.4. Classification of creams

- **By destination:** Creams for the consumer market; creams for the industrial secondary processing market (sauces, ready meals, desserts, etc.); creams for the industrial production of butter and its derivatives.

- **Depending on the fat content:** light cream (containing between 10 and 20 per cent fat); normal cream (containing at least 30 per cent fat).

- **Depending on treatment:** Raw cream (without any treatment); pasteurised cream; sterilised cream; UHT cream; frozen cream; deep-frozen cream.

- **Depending on how they are presented:** fluid cream; thick cream (matured); whipped creams (whipped by incorporating air); pressurised creams (packaged in airtight metal containers with nitrous oxide to ensure they are whipped); sweetened cream; flavoured creams (Figure 15).

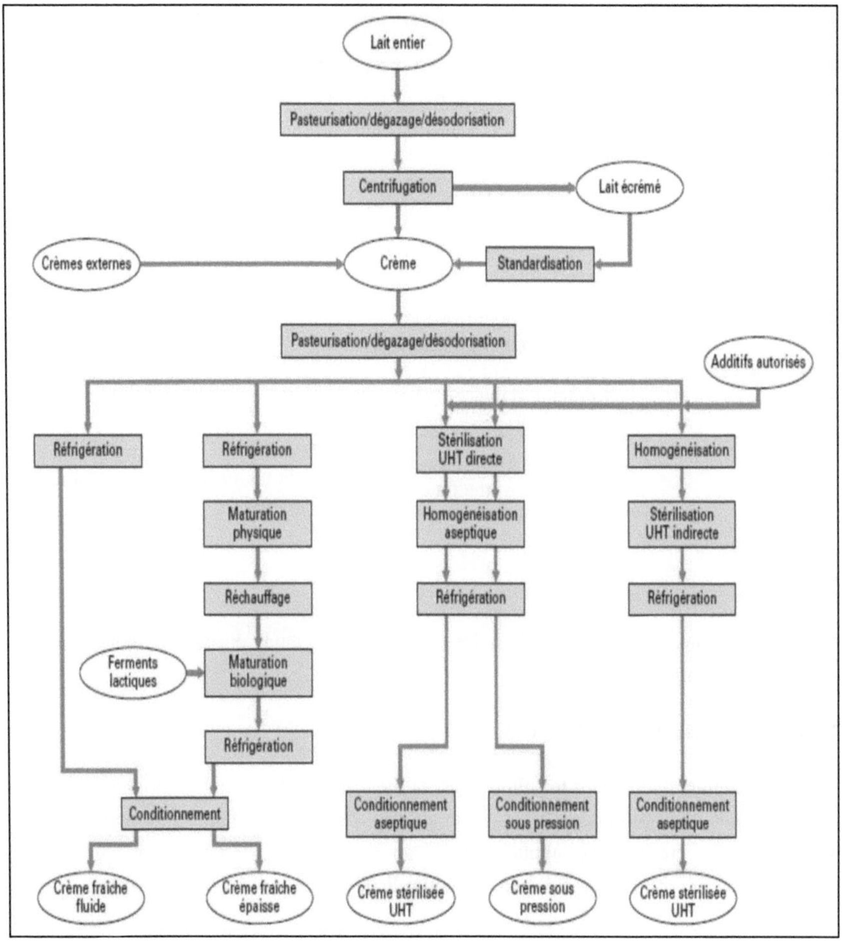

Figure 15: Production diagram for different consumer creams (Boutonnier, 2007).

2.1.5. Cream quality defects

Cream, like all dairy products, can have quality defects related to the raw material, the manufacturing processes or the preservation. Some of these defects are listed in the table below (Table 7).

Table 7: Defects in cream (Anonymous 1, 2019)

Defects	Origins	Prevention/correction
Upwelling (syneresis)	Excessive acidification. Cold chain problems during storage.	Fermentation, maturation temperature. T°C monitoring.
Cream too thick	Excess solid fat. Excess acidification.	Fat composition (summer/winter) Cold storage too long. Reduce biological ripening.
Butter in the cream	Excessive agitation.	Avoid shaking the cream too much.
Sour, acid taste	Excessive acidification. Contamination by spoilage bacteria. Problem in the cold chain during storage.	Fermentation, maturation temperature. Hygiene and packaging. T°C monitoring.
Rance	Oxidation. Alteration of the fat content.	Tightness of packaging. Milking machine and equipment hygiene.

2.2. Butter

Certain operations are identical to those for cream, namely standardisation, pasteurisation and deaeration/deodorisation.

2.2.1. Physical ripening of the cream

Physical ripening, which aims to solidify part of the triglycerides, is an essential operation for obtaining butter of optimum and consistent quality, despite the degree of variability in the quality of the cream. The application of an appropriate thermal cycle enables the crystallisation of the triglycerides to be controlled, thereby correcting the effects of the season.

Temperature cycles are chosen to optimise the liquid fat content, which should be between 65 and 90% of the total fat content. If the fat content is too high, the buttermilk loses fat, as the fat globules tend to be homogenised rather than agglomerated. On the other hand, if the fat content is too low, phase inversion will be delayed and the butter granules will not agglomerate.

Very rapid cooling causes the formation of mixed crystals in which low-melting triacylglycerols are trapped inside a structure of high-melting triacylglycerols. The resulting butter will be very hard. Slower cooling allows better control of fat crystallisation and a softer texture. The programme of temperature cycles (hot-cold-cold or cold-hot-cold) is chosen according to the melting point of the fat in the cream (given by the iodine value). The maturing time is usually 12 to 15 hours and it is also during this stage that ferments are added to produce a cultured butter, if required.

2.2.2. Biological ripening of the cream

This operation is carried out as part of traditional production processes. In industry, the Netherlands Institute for Dairy Research (NIZO) developed a method in 1976 which enabled manufacturers to do away with the biological maturing of cream. The cream only undergoes physical maturation. The sweet cream is transformed into butter, and acid and aromatic concentrates are injected directly into the butter, after phase inversion, so as to adjust the chemical composition of the butter in the butyrator.

There are three processes for making butter:

- Concentration process
- Combination process
- **Agglomeration process:** this is the process we will be focusing on here, as it is the most widespread in the world (industrial and small-scale processes). It has become established thanks to its control over the quality of the finished product, its flexibility of use and, above all, the productivity of the equipment it uses.

Butter is made in the following stages:

2.2.3. Churning

The cream can be churned in an old-fashioned or mechanical churn (Figures 16 a and b) or in a continuous industrial butyrinator (Figure 17). The churning temperature should be 10-12°C to facilitate phase inversion. Phase inversion can occur in 30 to 60 minutes in the case of the churn or in a few seconds in the case of the continuous process.

The churn is filled to a maximum of 50% of its capacity, because for phase inversion, it is essential to pass through a cream foam stage, in order to bring the fat globules together and facilitate their agglomeration. Under the action of repeated shocks, on the one hand air is incorporated and the fat globules migrate to the air/cream interface and on the other hand 20 to 30% of the fat globules are damaged. When the fat globules come into contact with the air bubbles and some of their membrane material is deposited on the surface of these bubbles, the globules adhere to the air/serum surface and release some of their liquid fat onto this surface. The eventual melting of the air bubbles gradually brings the fat globules closer together and causes them to agglomerate. As a result, more and more liquid fat, in the form of fine globules accompanied by membrane fragments, is released into the serum as the air-serum surface area decreases.

Around a third of the fat found in buttermilk is made up of finely dispersed triacylglycerols (<0.3 μm) that are very difficult to skim off by centrifugation. However, the presence of liquid fat is essential for phase inversion, as very high levels of solid fat tend to stabilise the foam by creating a rigid structure in which air bubbles are trapped (whipped cream).

At a certain stage in the manufacturing process, the gas pressure inside the foam's air bubbles is insufficient to support these agglomerates of fat, so the foam collapses - phase inversion. At this point, two phases appear, one solid, the grains of butter, and the other liquid, the buttermilk. Then, when the butter grains have reached the desired size, the rotation is stopped and the buttermilk is drained by gravity.

a: Mechanical butter churn (https://www.exportersindia.com/k-s-system/butter-churning-machine-5444518.htm);
b: Old-fashioned wooden churn https://www.alamyimages.fr/photos-images/baratte-%C3%A0-beurre.html)

Figure 16: Churn models.

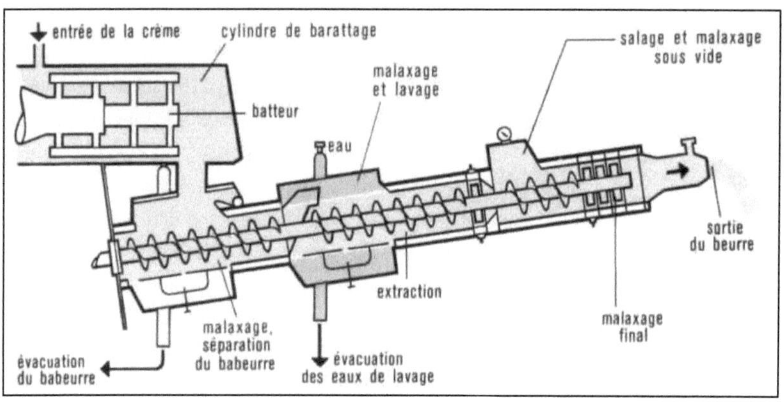

Figure 17: Diagram of a butyrator (longitudinal section)
http://www.larousse.fr/archives/grande-encyclopedie/page/7776

2.2.4. Washing the butter grains

Washing the butter grains eliminates the inter-granular buttermilk. This operation improves preservation because some of the micro-organisms are removed with the wash water.

The butter is washed in cold water to lower the temperature and facilitate the next step (mixing), then ice-cold water (2 to 6°C) is added in a volume equal to that of the buttermilk removed. Leave to stand for 5 to 10 minutes. Run the churn at slow speed to rinse the butter grains, then drain off the water. Two or three successive rinses can be carried out until the water remains clear after stirring.

2.2.5. Mixing

The butter is kneaded in the churn. This operation brings the grains of butter together into a homogeneous mass, eliminates the remaining buttermilk and excess water, ensures better preservation and improves the consistency of the butter. Kneading causes the temperature to rise. This is why the butter must be cooled when it is washed (desired temperature 9 to 10°C). The churn rotates at a slow speed. It is then emptied by gravity. One of the aims of kneading is to break up some of the GG intact in the butter grains. Butters rich in small GGs tend to have a greater number of intact GGs that cannot necessarily be reduced by kneading. These butters, compared with butters rich in large GGs, are moister. They are therefore more spreadable.

2.2.6. Butter structure

The fat globule plays a major role in butter production, and the physical and chemical characteristics of milk fat vary according to breed, lactation period and diet. In summer, for example, the proportion of unsaturated fatty acids makes the butter softer than in winter.

The fatty phase consists of a large number of intact GGs embedded in a liquid MG cement. Within the GGs and the liquid MG cement are crystals of solid MG. The arrangement and network that the MG crystals can form in the cement are responsible for the firmness of the butter. There is also a significant proportion of dispersed water in the form of small droplets (1-25 µm), which are generally spherical and never have a birefringent layer, and air bubbles (> 20 µm).

The main criterion for assessing the functional quality of butter is spreadability. It depends on the structure of the butter: the state of crystallisation of the fat linked to the GA composition (ratio between solid fat and liquid fat, and organisation and stability of the crystallised fat) and the humidity are the main factors to be taken into account.

2.2.7. Storage

The butter is packaged and stored in a cold room. This storage allows the crystalline structures to be rearranged. The descent to storage temperature and its maintenance for several days are responsible for the establishment of more stable crystalline structures. Butters rich in small GGs become harder and have a more homogeneous structure. However, these butters tend to melt over a more restricted range of temperatures.

2.2.8. Concentrated butters

The aim of these technologies is to eliminate virtually all the aqueous phase of the butter in order to obtain a product that is stable over time and can be stored at room temperature. Three technological approaches have been used. Two of them use cream (a fresh product) as the raw material. The result is a good quality finished product commonly known as "anhydrous milk fat" or "butter oil". The third method, using butter as the raw material, is hardly ever found industrially these days.

2.2.9. Butter quality

According to Codex Alimentarius, milk used in the manufacture of products covered by the provisions of this standard must comply with the maximum limits for contaminants and toxins prescribed for milk in the General Standard for Contaminants and Toxins in Products for Human and Animal Consumption and with the maximum limits for residues of veterinary drugs or pesticides prescribed for milk.

According to national legislation, butter is a fatty product derived exclusively from milk and products obtained from milk in the form of a water-fat emulsion. Butter contains a minimum of 82g of milk fat, a maximum of 2g of non-fat dry matter and a maximum of 16g of water per 100g of finished product. With the exception of raw butter, all butters are prepared from pasteurised milk fat. The peroxide value of butter is set at a maximum of 0.5 milliequivalents of active oxygen per kilogram of fat. The free fatty acid content is set at a maximum of 0.35%, expressed as oleic acid. The maximum concentrations of contaminants tolerated in butter are set as follows: Lead: 0.05mg/Kg; Iron: 2.0 mg/Kg; Copper: 0.05mg/Kg. The microbiological criteria applicable to pasteurised butter are set by the Official Journal. The butter must meet compositional and hygiene standards, which are checked by means of appropriate analyses. Yeast and mould counts provide information on the hygienic conditions under which the butter is manufactured: their presence, if any, is an indication of recontamination after pasteurisation of the cream. In addition, butter is subject to sensory quality standards assessed by means of a scoring scale following examination of flavour and texture.

- ***Butter defects***

The main flavour and texture defects in butter and the probable causes of their appearance are listed in Table 8.

Table 8: Flavour and texture defects in butter (Lapointe-Vignola, 2002).

	Defects	Probable causes
SAVEUR	Taste of food and absorption	Taste of certain foods and medicines consumed by the cow; strong-smelling ambient substances (sanitising agents).
	Fermentation flavour	Acid fermentation of cream; yeast and mould fermentation (old cream). Putrid flavour due to the growth of psychrotrophic proteolytic bacteria (Pseudomonas) cheese flavour.
	Oxidation flavour	Oxidation of phospholipids. Reminiscent of the flavour of paper, cardboard, wood and metal; oxidised fat (sweet flavour) due to the oxidation of triglycerides (less frequent).
	Taste of fish	Degradation of phospholipids, mainly lecithin
	Storage flavour	The fresh flavour diminishes slowly and gradually during storage due to the onset of auto-oxidation of the fats.
TEXTURE	Sticky texture adheres excessively to an object, leaves a trail on the probe or knife	Too high a proportion of liquid fat at room temperature, characteristic of winter butter, over-kneading.
	Brittle, crumbly texture (the fault of a butter that is too firm, lacks cohesion and is difficult to spread).	Especially in winter (higher proportion of firm fat) the fat crystallises into large crystals, leaving too little liquid fat to act as a binding agent.
	Weak texture	Especially in summer due to too high a proportion of liquid fat
	Greasy texture and ointment texture	These defects are the result of the over-mixing of soft and firm butter, respectively
	Floury texture	Coarse granular state caused by the presence of coarse crystals of demulsified fat formed, for example, as a result of partial churning following excessive agitation of the cold cream.

2.3. Traditional butter

Fresh butter or Zebda is obtained after churning the curdled milk (Rayeb), which is gradually increased with a quantity of lukewarm water (40 to 50°C) at the end of churning to encourage the fat globules to agglomerate and increase the butter yield. The lipid globules appear on the surface and are recovered at the end of the churning process. The fresh butter obtained has a soft consistency due to its high water concentration.

The process consists of leaving the milk to itself in an earthenware pot (*Raouaba*) until it coagulates. This takes place at room temperature and lasts 24 to 72 hours depending on the season.

The curds formed are stirred with a spoon or ladle to make them easier to transfer to the container used for churning. Since ancient times, three types of traditional churn have been used in Algeria, depending on the region. The Chaouias of the Aurès and the Saharan nomads use the Chekoua (Figure 18), made from the skin of a goat or sheep after painstaking processing. The rayeb-filled Chekoua is suspended from a tripod or beam, and vigorously shaken back and forth until the aggregates of fat particles fuse together. The fat particles then clump together to form grains of butter.

Figure 18 : *Chekoua*; Sheep/goat skin sheathing (https://www.twitter.com/photo-images/chekoua.html).

In the eastern Kabylie region, women used earthenware utensils called *Mezla* (Figure 19) or *Artoul* for small volumes. The *Mezla is* placed on a piece of cork and shaken to extract the butter. On the side of the *Mezla* is a small hole that is closed with a cloth called *Anfous*, which the women remove from time to time during the churning process to evacuate the air that forms inside the churn. The opening of the churn is hermetically sealed with a piece of goatskin called *Afzaz* tied around the main opening of the churn with a string.

Figure 19 : Mezla, clay churn (https://www.mangeonsbien.com/photo-images/mezla-rouaba.html)

On the heights of the Djurdjura, Kabyle women use *Thakhssayeth Oussendou*, also known as *Thakhchachet* (Figure 20). This choice was not made at random, given the steep, mountainous terrain of Kabylia, where a plant called calabash grows. This plant produces a fruit which, when ripe, becomes rigid and empty inside, and is used as a traditional churn in Kabylia.

The operator must shake vigorously with both hands. The churning operation lasts from 40 minutes to 1 hour 15 minutes. The zebda is generally collected by hand.

Figure 20: Thkhssayeth Oussendou or Thakhchachet; Churning calabash (https://www.nissa22.unblog.fr/photo/chekoua-kabyle.html).

2.4. D'han or Smen

Algerian Dhan or Smen is an ethnic dairy product which is a traditional fermented butter made from whole raw milk using empirical methods. This product is an important part of the Algerian diet and represents a gastronomic heritage that must be preserved and protected. Depending on the region and culinary habits, raw milk (cow's, goat's and/or sheep's milk) is fermented spontaneously until it coagulates. The resulting coagulum (Raib) is then churned using various tools to obtain the butter used to prepare Smen Dhan. Two different preparation methods are then used, depending on the region studied. Households in Batna, Khenchela and Sétif salt the butter directly, while in the regions of Biskra, Jijel, El Oued and Ouargla, a different process is used which first involves heat-treating the butter, followed by the addition of one or more ingredients. After packaging in a traditional ceramic vessel (*Ezzir*) and other containers, a maturing stage is applied, lasting from one month to several years.

2.5. Clarified butter

It is generally prepared directly by the cook by melting the butter, which gives rise to three phases. Only the intermediate phase, a clear yellow butter oil, is recovered after decanting and removing the supernatant (destabilised proteins) and the heavy fraction (aqueous phase). This oil, which solidifies on cooling, is used as a cooking fat, with the advantage that it does not blacken or splatter during heating. In addition to these culinary uses, which are more common in the catering industry, this product is also available in jars in some European countries, and even in sheets, as in France, where it has recently been relaunched under the commercial name of 'cooking butter'.

2.6. Other craft products

- **Tunisia**

In southern Tunisia, two products are produced from the spontaneous fermentation of milk: the first is rich in protein and is known locally as 'Leben'. The second is rich in fat and is known locally as "traditional butter" or "Zebda beldi".

Butter oil (called sman) is obtained by heating traditional butter and separating the fat (butter oil) from the whey. A large proportion of butter oil is used in cooking.

- **Morocco**

Traditional dairy products are highly valued by consumers for their nutritional properties and organoleptic characteristics. Among these products, "Smen" is a fermented and salted Moroccan butter consumed mainly as a food additive and as an ingredient in processed foods. It is an artisanal product specially made to protect and preserve butter for a long time, and is used as a fundamental food ingredient. Its unique and pleasant aroma can enhance the taste and flavour of certain Moroccan dishes.

- **Jordan**

Traditionally, Jordanian butter has always been made from yoghurt. Once a sufficient quantity of sheep's milk has been collected, it is fermented and then churned by mechanical agitation until butter granules are formed. In making butter from yoghurt, the aim of churning is to extract as much fat as possible by transferring the oil-in-water emulsion to the water-in-oil emulsion. The liquid that remains after butter extraction "the buttermilk" is used to produce a type of dry yoghurt called Jameed, a local dairy product in Jordan.

- **Egypt**

Cooking butter is one of the most popular types of fat consumed in Egyptian homes. It is consumed as butter, used as oil for food preparation or for cooking. Cooking butter is obtained by churning the mechanically separated cream, after storage in a skin bag called a "Kerbaî" made from a one-piece goatskin for 2 to 3 days at room temperature. An excessive amount of sodium chloride is added to the outer surface of "Kerba" during storage. Churning is carried out after hanging the "Kerbaî", which is filled with cream, and shaking it vigorously back and forth until the fat globules coalesce, which is indicated by the sound of the butter pieces during shaking. After churning, the butter is processed by hand and stored in refrigerators until it is sold.

- **Ethiopia**

In Ethiopia, there are two types of butter, ripened / rancid and fresh, known locally as besalkibe and lega kibe, respectively. However, another study reported that there are three types of butter in Ethiopia, namely lega, mekakelegna and besal, which refer to fresh, semi-ripe and rancid butter respectively, depending on the degree of lipolysis of the butter. Butter is made and processed exclusively by the women of each community in Ethiopia.

Locally produced butter is semi-solid at room temperature. It has a pleasant smell when fresh, but with increasing storage time, changes will occur in smell and taste, unless it is refrigerated or made into traditional ghee (dhadhabaksaa / nitirkibe) by boiling it with spices.

- **India**

Milk can be used in the production of products such as dahi (traditional Indian fermented milk), lassi (traditional Indian buttermilk with a thick consistency), chhach (traditional Indian buttermilk with a thin consistency), makkhan, ghee and so on.

Makkhan is a high-fat dairy product traditionally grown in India and Pakistan. It is quite similar to western butter. Whole milk or cream, whether processed or not, is first converted into dahi and then churned into makkhan. When makkhan is used in the production of ghee, it also adds a distinct and pleasant aromatic flavour to the ghee.

2.7. Quality of craft products

Traditional butter is not subject to any regulations. It is made from the milk of different animal species (cow, sheep, goat, etc.), and the production processes vary from country to country and sometimes from region to region within the same country. It is therefore clear that its quality varies according to these factors.

In this chapter, we will report on the qualities of butter according to the studies carried out.

- **Algerian**

In the Jijel region of eastern Algeria, a study revealed the microbiological and physico-chemical properties and fatty acid composition of a traditional butter produced from cow's milk. The presence of lactic acid and psychrotrophic bacteria as well as yeasts was revealed, while staphylococci or lipolytic bacteria were not detected. Significant differences were found in the chemical values among the samples. pH values ranged from pH 4.64 to pH 5.53. Moisture and impurities exceeded 17.5% and 9.19% respectively. Acid value, peroxide value, saponification value and iodine value ranged from : 23.56-31.35 mg KOH/g, 1.6-4 meq/kg, 140.25-228.60 mg KOH/g and 35.35-53.69 mgI/100g respectively. Finally, the fatty acid composition showed that palmitic acid and oleic acid were the main saturated and unsaturated fatty acids.

- **Tunisian**

Physico-chemical characterisation showed a fat content of less than 80% and a high water activity value. The saturated fatty acid content was higher (71.84%) than that of unsaturated acids (27.09%). The main fatty acids in the butter samples were myristic, palmitic, stearic and oleic acids. During storage at 4 and 10°C, pH decreased and titratable acidity increased. The number of lactic acid bacteria showed relatively small changes during storage at 4 and 10°C, while the number of yeasts and moulds increased at all temperatures. The study also showed the effect of heating (60°C) on certain quality characteristics (absorption at 232 and 270 nm, peroxide value, free fatty acid content, viscosity, texture, colour and fatty acid composition) of traditional Tunisian butter oil (TTBO). The results showed that TTBO was resistant to oxidation. All these characteristics support the incorporation of TTB in food formulation.

- **Iranian**

Comparison of the physico-chemical characteristics of Traditional Butter (TB) with European (ES) and Iranian National (NS) standards showed that the salt content, iodine value and acidity of all samples were consistent with ES. The peroxide value, moisture and solids non-fat content of some butter samples were higher and the fat content of 74% of them was lower than the ES values. The saponification value of 26% of the samples was outside the ES range.

Microbiological qualities revealed that the number of coliform bacteria and psychrotrophic bacteria was higher than that of NS in 8% and 4% of samples, respectively. Unlike NS, *E. coli* and coagulase-positive staphylococci were detected in 10% of samples. The number of moulds and yeasts in all samples was lower than the NS value, and the organoleptic property score in all samples was higher than 8 points.

The study therefore concluded that despite the high quality and low microbial contamination of the butter samples, greater supervision of the production process for traditional products is recommended.

- **Egyptian**

Cooking butter in the governorate of Beni-Suef was examined for psychrotrophic bacteria, total coliforms, faecal coliforms, moulds and yeasts. In addition, pathogenic bacteria such as *E.coli*, *S.aureus* and *Ps.aeruginosa* were also examined.

Microbiological examination revealed that 100, 100, 36.7, 31.7, 31.7 and 23.3% of the samples examined were contaminated with psychrotrophic bacteria, moulds and yeasts, coliforms, faecal coliforms, *E.coli* and *S.aureus*, respectively. None of the cooked butter samples examined contained *Ps.aeruginosa*.

The mean values for sodium chloride and titratable acidity were $0.57 \pm 0.05\%$ and $0.20 \pm 0.013\%$, respectively. The study showed that cooking butter is produced under unhygienic conditions and without good manufacturing practice.

- **Sudanese**

A survey was carried out to study the microbiological quality of butter produced in the Khartoum region. The study compared several types of butter (traditional, made in a dairy and made in the laboratory). Microbiological characteristics were assessed at time intervals (days) for samples stored at 5.0 ± 1.0°C, and at 25 ± 2.0°C. The results showed that the total number of viable bacteria (TVBC) increased significantly from the beginning of the storage period to the end in butter from all treatments, the increase being remarkable for samples stored at 25°C. Coliform bacteria increased to a lesser extent compared to TVBC, the increase being crucial in samples stored at 25°C. Storage temperature significantly affected counts of yeasts and moulds, *Pseudomonas aeruginosa*, proteolytic and lipolytic bacteria, however, the difference in lipolytic bacteria counts between the two temperatures was not significant. The study concluded that the microbiological quality of butter is mainly affected by the way the butter is manufactured and the storage conditions during the storage period.

- **Ethiopian**

A survey was carried out to access the physicochemical and microbial quality of butter from the Menz district. Overall values ranged from 2.09% to 83.96% for moisture. An overall average of 3.94×10^9 , 2.66×10^6 , 1.83×10^6 , for the total number of aerobic mesophilic bacteria, the total number of coliforms and yeasts and moulds were observed in the samples from farmers and $3,44 \times 10^9$, 3.03×10^6 , 1.31×10^6 and 3.26×10^9 , 1.61×10^6 , 1.77×10^6

for the samples collected from traders and for the butter manufactured by the investigators, respectively. For the samples collected in Tarmaber and Addis Ababa, these values were 4.19×10^{9} , 2.69×10^{6} , 1.56×10^{6} and 4.20×10^{9} , 2.10×10^{6} , 1.45×10^{6} respectively. The study concluded that butter production and processing are unhygienic in the study areas and that the traditional production system is substandard.

- **Malaysian**

A study of the microbiological, physico-chemical and fatty acid structure characteristics of a commercial cow's butter (BCC) sold in Malaysia showed the existence of variations in physico-chemical parameters between butter samples. The pH range was between 3.32 and 4.90. The water content of all but two samples was within the international standard range. The peroxide and iodine values of all samples were within the international standard limit. Determination of fatty acid composition showed the prevalence of saturated fatty acids dominated by palmitic acid, with a low level of unsaturated fatty acids, dominated by oleic acid. The results also showed the minimal presence of total aerobic mesophilic bacteria (TAB) and psychrotrophic bacteria, while coliform bacteria were not detected. Moulds and yeasts were detected in all samples at minimum count. As a result, the microbiological quality of commercial butter samples was found to be generally good.

- ***"Moroccan "Smen***

Moroccan 'Smen' is produced by enzymatic and chemical reactions. Lipolysis is considered to be the main biochemical pathway leading to organoleptic changes during the maturation of "Smen". Lipolysis can result mainly from microbial lipases. The release of free fatty acids (FFA) during the ripening period progressively inhibits *Leuconostoc* and *Lactococcus* bacteria, while *Lactobacillus* take over because of their ability to metabolise FFA. These reactions have a major effect on the taste, aroma and organoleptic characteristics of 'Smen'. Short-chain (C4-C10), unsaturated (mainly oleic acid (C18:1)) and saturated (C12-C20) free fatty acids are respectively released into the product as a function of maturation time. In addition, the chemical oxidation of milk lipids can make a lesser contribution to the flavour of "Smen" and can lead to unpleasant odours.

- **"Jordanian *"Ghee", "Jammed" and "Smen Baldi***

The traditional production process for 'Jameed' and 'ghee' in Jordan is primitive and time-consuming. In addition, many challenges can affect the quality of the products, such as poor sanitation, poor microbiological quality of the milk due to poor hygiene during milking, lack of effective cooling of the milk and lack of pasteurisation of the milk. Plastic and aluminium containers are generally used in all stages of processing, both of which are unsuitable for food production. In addition, the lack of cooling at the milk collection stage reduces the quality of the milk due to the growth of micro-organisms, leading to the development of acidity and the production of numerous materials that seriously affect human health and the production process. Accelerated enzymatic activity due to the high storage temperature leads to degradation of fats, resulting in rancidity of the final products, particularly ghee. While protein degradation gives rise to an undesirable taste in 'Jameed'.

A survey was carried out to assess the oxidative quality under traditional "Samen Baladi" production conditions and the quality factors under controlled conditions of good manufacturing practice and laboratory temperature. The overall results concluded that all the Samen *Baladi* samples studied failed to meet Jordanian regulations and *Codex* standards for peroxide value and refractive index, as well as the percentage of free fatty acids. The study concluded that it would be highly advisable to find a structural approach to controlling peroxide formation in 'Samen Baladi', as a conventional method currently adopted requires heat treatment.

- **"Traditional *ghee* in Uganda**

A study compared the quality of butter/'ghee' made using traditional churning (gourds and calabashes) and mechanised churning at four locations, as well as a 'control' butter sample made under laboratory conditions following a standard procedure. The samples were analysed for microbial safety, type and concentration of free fatty acids and sensory attributes. Total viable count (TVC), total coliform count (TC), *Staphylococcus aureus*, *Salmonella*, yeast and mould were determined using International Standards Organisation (ISO) standards. The fatty acid profile was determined by gas chromatography. A sensory evaluation of the aroma, smell, taste, mouthfeel and overall acceptability of the products was also carried out. Total viable counts in all samples ranged from 10^2 to 10^7 cfu/g. Total coliforms were detected in samples from Kiboga between 10 -10^{13} cfu/g, while none were detected in samples from other regions. Yeasts and moulds were detected in the range 10 -10^{25} cfu/g. *Staphylococcus aureus* was only detected in butter samples from the Kiboga region (10^2 cfu/g) while *Salmonella* was not detected in any of the samples. The fatty acid profile consisted of saturated fatty acids, monounsaturated fatty acids, polyunsaturated fatty acids, trans fatty acids, omega-3 fatty acids, omega-6 fatty acids and omega-9 fatty acids. Saturated fatty acids were the most dominant in the butter and ghee samples, ranging from 70 to 82%, while trans fatty acids were present in lower concentrations. From the point of view of overall acceptability, butter/'ghee' manufactured using traditional churning and the mechanised system obtained the highest score. However, no significant difference was shown in the organoleptic parameters analysed in all samples ($p>0.05$). Therefore, butter/"ghee" produced using mechanised churning is as acceptable and as safe microbiologically as butter/"ghee" produced using traditional churning.

PROTEIN PROCESSING

The transformation of milk protein is a complex process that can occur in a variety of contexts, such as the manufacture of dairy products, fermented milks, yoghurt and cheese. It is the caseins that have the ability to coagulate when subjected to specific conditions, such as a drop in pH during fermentation or the addition of coagulating enzymes for the production of these foods.

1. Caseins

There are four types of casein naturally present in milk:α s1,α s2, β-casein and κ-casein. γ-caseins are peptide fragments resulting from the degradation of β-casein by plasmin. The 4 main caseins are present in milk in proportions of 37, 10, 35 and 12% respectively, and have their own physicochemical characteristics.

Caseins have an isoelectric point (pHi) of 4.6 at 20°C and precipitate at this pH, but they are not sensitive to temperature variations.

They are proteins rich in lysines and glutamyl and aspartyl residues. Molecules between 20-24000 Da in size. Caseins are strongly hydrophobic in the order β > κ > αs1 > αs2.

Casein αs1 is made up of 199 amino acids. It is negatively charged (-20.6 at pH 6.6). It has 3 hydrophobic regions. It aggregates and precipitates at relatively low calcium concentrations.

Casein αs2 contains 207 amino acids. It is the most hydrophilic of the caseins (three hydrophilic zones). These 3 zones are the preferred sites for calcium phosphate fixation and are responsible for the high sensitivity of this casein to calcium, which is potentially higher than that of αs1 casein.

β-casein, the most hydrophobic of the caseins, is made up of 209 amino acids. In view of its hydrophobic nature, β-casein has the particularity of having a temperature-dependent behaviour. Indeed, lowering the temperature of the milk (4°C) induces solubilisation of this casein in the soluble phase.

κ-casein, which is not very sensitive to calcium, is made up of 169 amino acids. This casein has a mixed hydrophobic and hydrophilic (amphipolar) character.

1.1 Casein micelle

To date, the structure of the casein micelle is still debated and therefore not yet fully established. The micelle is a complex, dynamic system that is difficult to observe under conditions that are representative of the milk's native state, and evolves as a function of the physicochemical conditions applied (variation in temperature, acidification, alkalinisation, etc.).

Casein micelles are quasi-spherical particles ranging in size from 50 to 600 nm. They are colloidal structures formed by milk caseins and phosphate and calcium in the form of small critalline entities. At 37°C and physiological pH (pH 6.7), casein micelles contain almost all the caseins present in milk, as well as a high proportion of calcium (around 2/3) and inorganic phosphate (around ½). The molecular weight of casein micelles varies from 10^6 to 10^9 Dalton (about 10^8 Dalton on average). In addition, it is important to note that casein micelles are assemblies made up of αS1, αS2, β and κ caseins in approximate proportions of 4/1/4/1. Mineral elements, mainly calcium and phosphate, represent approximately 7% of the dry mass of casein micelles.

The core of the micelle is formed by αs and β caseins linked by essentially hydrophobic interactions. Aggregation of these polymers with calcium phosphate leads to the formation of a macropolymer whose surface is bounded by κ-casein. Caseins play a central role in the transformation of milk into cheese and most other dairy products, and their structural and physical properties have been the subject of research for many years.

Based on the chemical, biochemical and physical properties of casein micelles and individual caseins, the most widely discussed models proposed are the sub-micellar model and the open structure model.

- **Sub-micellar model**

This model, based on various previous studies, characterises casein micelles as agglomerations of protein subunits called sub-micelles, measuring between 10 and 20 nm (Figure 21). The sub-micelles contain varying proportions of the different caseins αs, β and κ. Their association within the micelles is facilitated by calcium phosphate. The sub-micelles are composed of :

- a core mainly composed of αs caseins, which is hydrophobic

- a hydrophilic outer layer, where the κ-casein forms a sort of "hair", which maintains the stability of the colloidal suspension thanks to its high negative charge and steric repulsion. The polar parts of the different caseins facilitate their binding via calcium phosphate clusters. The sub-micelles richest in κ-casein are found mainly on the surface of the micelles.

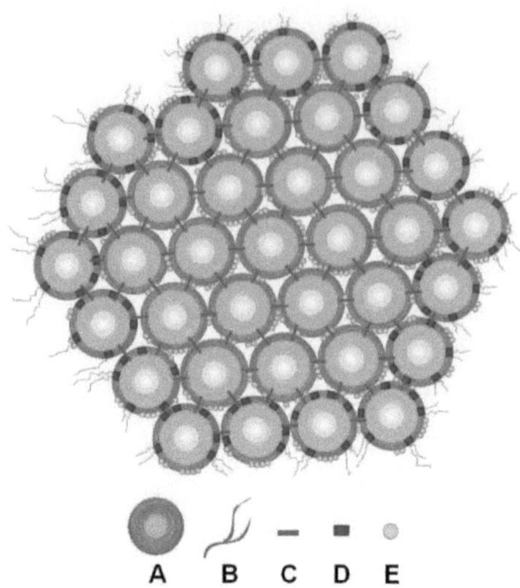

A: submicelle; B: protruding chain; C: calcium phosphate;D: kappa casein; E: phosphate group

Figure 21: Submicellar model of casein micelles
https://www.sciencelearn.org.nz/images/952-casein-micelles

- **Open model**

In this model, which has evolved, casein micelles are described as spherical, highly hydrated particles with an open structure. This network structure model is accepted because it is consistent with the dynamic view and equilibria with the aqueous phase of milk, where calcium phosphate is distributed in the micelle. Hydrophobic and electrostatic interactions between the caseins hold the micelles together, while the nanoclusters act as anchor points for the caseins through their phosphoserines, facilitating the growth of size-limited casein polymers. The κ-casein is localised on the surface of the micelle, ensuring its stability through steric and electrostatic repulsions that prevent natural aggregation. On average, a medium-sized casein micelle contains around 800 nanoclusters of amorphous calcium phosphate, with a diameter of around 100 nm (Figure 22).

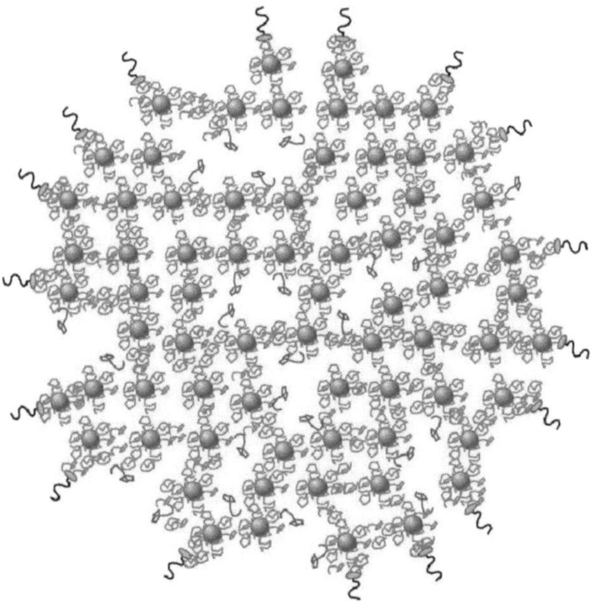

The α- and β-caseins (orange) interact electrostatically with the calcium-phosphate nanoclusters (grey) to stabilise the inner network. Some β-caseins (blue) interact via hydrophobic interactions with other caseins. The κ-casein is located at the surface of the micelle (green), with its hydrophilic caseinomacropeptide extended outwards (black).

Figure 22: Schematic representation of the proposed structure of casein micelle (CN) in milk (Dalgleish and Corredig, 2012).

2. Coagulation process

Milk coagulation corresponds to a destabilisation of the original micellar state of the milk caseins. This destabilisation can be achieved in two ways:

o or by fermentation or acid coagulation using lactic bacteria contained in the milk's indigenous flora and/or supplied in the form of ferments,

o by enzymatic means using coagulating enzymes, in particular rennet

2.1. Acid coagulation

2.1.1. Lactic flora

Lactic acid bacteria are aero-anaerobic or micro-aerophilic. They are also demanding bacteria from a nutritional point of view, as they are unable to synthesise a number of amino acids.

Their inability to synthesise the amino acids they need for growth means that lactic acid bacteria have to find these molecules in their environment via a process of nitrogen nutrition. Cultures or associations of strains of different lactic acid bacteria can encourage symbiosis, as in yoghurt. Their culture requires media rich in sugars, amino acids, fatty acids, salts and vitamins and low in oxygen, and they are able to live in very acidic environments (pH of 4 to 5). The main genera of lactic acid bacteria are : *Lactobacillus, Leuconostoc, Lactobacoccus, Streptococcus, Pediococcus, Carnobacterium, Enterococcus, Oenococcus, Tetragenococcus, vagococcus and Bifidobocterium.* Table 9 shows the different genera of lactic acid bacteria of interest in food microbiology and their main characteristics and habitats.

Table 9: The different genera of lactic acid bacteria (Federighi, 2005).

Type	**Morphology and characteristics**	**Fermentation**	**Main habitats**
Lactobacillus	Bacilli, Thermophilus or Mesophilus	Homofermentative or heterofermentative	Human, dairy, meat and vegetable products
Carnobacterium	Psychotropic bacilli, not very acid-tolerant	Hetero-fermentary	Meat, fish and dairy products
Lactococcus	Shells, Mesophilic.	Homofermentation	Dairy products, vegetables
Streptococcus	Thermophilic shell	Homofermentation	Dairy products
Enterococcus	Shells, Mesophilic.	heterofermentative	Human and animal intestines, dairy products
Pediococcus	Shell, Halotolerant mesophiles	Homofermentation	Beer, vegetable products, sausages
Tetragencoccus	Tetrad shell, Mesophilic, halophilic	Homofermentation	Brine
Leuconostoc	Shells, Mesophilic	Heterofermentative	Plant products, dairy products
Oenococcus	Shells, Mesophilic	Heterofermontaires	Wine
Bifidobacterium	Irregular shape, Mesophilic	Acetic and lactic acid	Intestine of humans and animals
Vagococcus	Mesophilic shells, mobiles	Homofermentation	Human and animal intestines, dairy products

Based on the route taken and the end product of fermentation, lactic acid bacteria are divided into two groups:

o **Homofermentative processes** use glycolysis in its entirety, from lactose to pyruvate to lactate. To qualify as homolactic, this pathway must convert at least 90% of the lactose consumed into lactate. The homofermentative pathway is generally associated with bacteria of the genera Streptococcus, Lactococcus, Pediococcus and Lactobacillus (Figure 23).

o **In** addition to lactic acid, **heterofermentative bacteria** produce significant quantities of CO2 and ethanol or acetate. Some Leuconostoc and Lactobacillus bacteria take this route (Figure 23).

A third form of metabolism, known as the difidic fermentative pathway, is used by bacteria of the genus Bifidobacterium. Bifidobacterium have a particular pathway that produces mainly lactic and acetic acid.

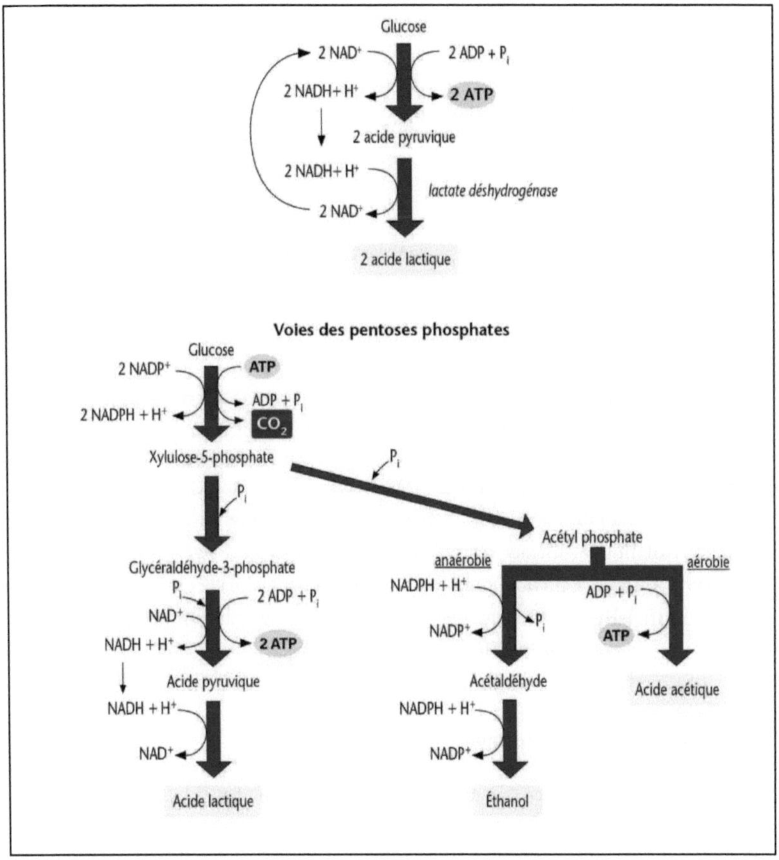

Figure 23: Illustration of homolactic fermentation (pyruvic acid is reduced to lactic acid) or heterolactic fermentation (using the pentose phosphate pathway) depending on the bacteria involved (Tessier, 2018).

2.1.2. Lactic coagulation phenomenon

The mechanism of fermentative coagulation, also known as acid coagulation, is electrochemical and induced by lactic ferments.

The main function of these bacteria is to break down lactose to produce lactic acid. Lactic acid is released as the micro-organisms grow and gradually neutralises the electronegative charges of the κ-caseins. The electrostatic repulsion between the casein micelles decreases as the medium becomes enriched with H^+ ions, and then disappears, causing the casein micelles to come closer together and aggregate.

2.2 Enzymatic coagulation

2.2.1. Enzymes

There are different families of enzymes used to coagulate milk: rennet and other preparations of animal origin, and plant, fungal or fermentative coagulants.

Rennet, composed of chymosin and pepsin, is obtained by maceration of rennets (the fourth stomach pouch of ruminants) from young ruminants fed on milk.

The recovery of proteolytic enzymes (chymosin and pepsin) as well as peptides, amino acids and other enzymes plays a role in the quality and typicality of cheeses.

Fungal enzymes, today essentially those of *Rhizomucor miehei* and *Cryphonectria parasitica.* Plant coagulants are traditionally used in many countries, mainly for domestic purposes, such as macerations of thistle flowers (cardosin enzyme), pineapple stems (bromelain), papaya leaves (papain), fig tree sap (ficin) and melon seeds (cucumisin). Fermentative chymosin is obtained by artificially introducing the prochymosin gene into microorganisms such as *Aspergillus niger var awamoris, Kluyveromyces lactis* and *Escherichia coli.* Fermentative chymosin is currently the most widely used coagulant in the world.

2.2.2. Enzymatic coagulation phenomenon

The mechanism of enzymatic coagulation is described in three phases.

- **A primary** or enzymatic **phase** during which the casein is hydrolysed. The remaining casein fragment has hydrophobic properties. In the first few minutes following the addition of the coagulating enzyme to the milk, the viscosity of the milk decreases. This is due to the reduction in the average size of the micelles following their hydrolysis.

- **A secondary phase** is when the micelles begin to come together. This phase begins when approximately 85 to 90% of the κ-caseins have been hydrolysed. The casein micelles then lose their affinity for the aqueous phase and will come closer together and aggregate under the effect of hydrophobic interactions, causing the milk to curdle.

- **The tertiary phase** is what is known as the gel cross-linking phase. The gel becomes increasingly organised and structured. At the microscopic level, we observe an increase in the bonds between the modified micelles, mainly hydrophobic and electrostatic interactions, as well as the formation of phosphocalcic bridges. On a macroscopic level, this corresponds to the hardening of the gel.

3. Transformations

3.1. FERMENTED MILKS

According to Codex, fermented milk is a dairy product obtained by the fermentation of milk, which may have been manufactured from products obtained from milk with or without modification of composition, by the action of appropriate micro-organisms and resulting in the reduction of pH with or without coagulation. These sourdoughs (micro-organisms) must be viable, active and abundant in the product at the date of minimum durability. If the product undergoes heat treatment after fermentation, the requirement relating to the viability of the micro-organisms no longer applies. Concentrated fermented milk is fermented milk whose protein content has been increased before or after fermentation to a minimum of 5.6%. The coagulation of fermented milk must not be obtained by any means other than those resulting from the activity of the micro-organisms used.

A wide variety of fermented milks are produced throughout the world: acidic fermented milks (Greek yoghurt, labneh), acidic and slightly alcoholic products (kefir, koumiss) and low-acid fermented milks (buttermilk, ribot milk). Some fermented milks contain micro-organisms called probiotics.

In this chapter, we will describe the fermented milks most commonly consumed in Algeria, namely: curdled milk or "Raib", whey or "l'ben" and yoghurt.

3.1.1. Raib and L'ben or Leben

Raib and ben are dairy products that have long been known in both rural and urban areas. They are generally prepared from raw milk using traditional methods. The production processes for leben and Raib (curdled milk) are similar, except that leben is subjected to mechanical agitation after incubation. It is important to clarify the difference between fermented milk and yoghurt:

- The lactic ferment contains mainly mesophilic bacteria for fermented milk and thermophilic bacteria for yoghurt.
- There is no enrichment in the production of fermented milk.
- Neither fruit nor flavourings are used in the production of fermented milk.

The traditional preparation of l'ben begins with the coagulation of the milk (cow's, sheep's or goat's). The raw milk is left at room temperature until it coagulates spontaneously into Raib (curdled milk) (for 24 to 72 hours, depending on the season). Skimming is carried out by churning (the traditional butter-making stage), leaving a thick liquid called lben (buttermilk or whey) inside. Finally, a pinch of salt is added, which is said to give the leben an aftertaste, but this is not common to all regions.

- **Leben Industriel (manufacturing technology)**

According to the FAO, acidified milk such as Leben or Lben is prepared using milk that is usually partially or fully skimmed. In countries where milk production is low, reconstituted milk is often used. After pasteurisation and any necessary degassing, the milk is cooled to 20-22°C and inoculated with 2.5 to 3% of a mesophilic lactic acid bacteria culture.

Fermentation continues for around 18 to 20 hours until coagulation and an acidity of 0.65 to 0.70 per cent lactic acid (65 to 70° Dornic). The curd is then more or less finely divided and stirred at the same time as it is cooled to 4-5°C. It is then packaged for sale or sold in bulk. When chilled, this slightly acidic, pleasant-tasting product can be kept for a week. It can be made with milk from various species (ewe's, goat's). Fermentation is generally carried out with mesophilic lactic acid bacteria, but can also be done with thermophilic bacteria; certain strains are prized for their ability to make the product viscous and stringy.

- **Quality of fermented milk**

At industrial level

-Milk quality control: Total germ count, somatic cell count and titratable acidity provide information about the milk's health quality. Analysis of the milk's fat and nitrogen content and screening for antibiotics help to assess its technological quality.

-Powder quality control microbiological quality

For microbiological criteria of finished products, according to legislation concerning pathogens (coagulase + *Staphylococcus*, *Salmonella* and *Listeria monocytogenes*)

3.1.2. Yoghurts

According to the Journal Officiel de la République Algérienne (JORA), yoghurt is a coagulated milk product obtained by lactic fermentation through the development of specific thermophilic lactic bacteria known as *Lactobacillus bulgaricus* and *Streptococcus thermophilus*, from milk and the following milk products: pasteurised milk, reconstituted or recombined pasteurised milk, whether or not skimmed, concentrated milk or dry milk, whether or not skimmed, or pasteurised cream or a mixture of two or more of these products. The incorporation, as a substitute product, of non-dairy fats and/or proteins is prohibited.

The specific thermophilic lactic acid bacteria must be inoculated simultaneously and must be present in the finished product at a rate of at least 10 million bacteria per gram in relation to the milky part. When released for consumption, the quantity of free lactic acid contained in the yoghurt must not be less than 0.8 grams per 100 grams of product.

Associative growth of yoghurt bacteria

The combined use of *S. thermophilus* and *L. bulgaricus* makes it possible to exploit the positive indirect interaction between these two species. This interaction, known as protocooperation, initially results in an increase in acidification rates. The stimulation of *S. thermophilus* by *L. bulgaricus* is achieved through the proteolytic activity of the lactobacillus, which releases small peptides and amino acids for the benefit of the streptococcus. In return, *S. thermophilus* supplies formic acid and CO_2 , both of which stimulate the growth of *L. bulgaricus*.

3.1.2.1. Manufacturing principles

There are three main stages in the yoghurt production process: preparation of the milk, fermentation and post-fermentation treatment of the product. The production diagram differs according to the type of product. There are two types of yoghurt:

- firm yoghurt, fermented after packaging in pots

- stirred yoghurt, in which fermentation takes place in a vat; the coagulum obtained is then dilacated and stirred to make it more or less viscous, then packaged in pots.

a. Reception and storage of raw milk or milk preparation

The raw material may be fresh milk, recombined milk (from skimmed milk powder and anhydrous milk fat), reconstituted milk (from skimmed milk powder) or a mixture. In all cases, it must be of good microbiological quality, free from antibiotics or other inhibitors and perfectly homogenised.

b. Standardisation

The milk must be standardised in terms of fat content, enriched with protein, and possibly sweetened, to meet the nutritional and organoleptic specifications of the products.

- **Fat standardisation** The fat content of commercial yoghurts ranges from less than 1% for low-fat yoghurts to 3.5% for whole-milk yoghurts. The skimmed milk and cream are then blended to achieve the desired fat content. The fat adds creaminess, masks acidity and enhances flavour.

- **Protein enrichment** Fat-standardised milk must be enriched with milk proteins to form a consistent, syneresis-free yoghurt. The quantities of protein added are variable and depend on the texture required (drinking yoghurt, firm yoghurt, stirred yoghurt). Final protein levels are between 3.2 and 5%, thanks to the addition of condensed milk, skimmed milk powder or milk replacer. Proteins improve texture and mask acidity

- **Sugar may be added:** 5-10% sugar may be added to the milk before fermentation. Sometimes sugar is added in two parts, one before fermentation and one after, so as not to slow down acidification. Sugar generally consists of sucrose, either crystallised or in liquid form (syrup). In the case of low-fat products, sweeteners (aspartam) are added. As these products are sensitive to heating, they are always added after the heat treatment.

c. Homogenisation

Fat-standardised, protein-enriched milk, which may be sweetened, forms the production mix. It is homogenised to reduce the size of the fat globules. This is essential to prevent the fat from rising during fermentation. Homogenisation also increases the viscosity of the yoghurt and reduces serum exudation (or syneresis) during storage of firm yoghurt. Finally, it gives the milk, and therefore the yoghurt, a whiter appearance.

d. Heat treatment and cooling

The mix undergoes heat treatment. Its purpose is to :

- destroy pathogenic micro-organisms that may be present and most common flora. It also removes any natural inhibitors.

- denature a large proportion of the soluble proteins, which increases the yoghurt's water retention capacity and allows these proteins to bind to the casein. The curd is firmer, and the tendency for serum to be expelled during storage is reduced.

Two systems can be used to treat the mix: batch or continuous treatment. Batch treatment is carried out in jacketed tanks by injecting steam directly into the jacket, generally at 85-90°C for 15-30 minutes. The continuous system involves the use of shell-and-tube heat exchangers or, more frequently, plate heat exchangers. Heat treatment rates vary depending on the installation: 30 min at 85°C, 5 min at 90-95°C, or 3 s at 115°C. The most common treatment is heating to 92°C for 5 to 7 minutes, with a circulation rate of between 4,000 and 20,000 l/h. The milk is then cooled to fermentation temperature.

e. Milk fermentation

Ferments are marketed in frozen or freeze-dried form. They are added, in sufficient quantity (for example between 5.10^5 and 5.10^6 UFC/ml), which corresponds to an inoculation rate of between 2.5 g and 70 g per 100 litres of milk depending on the bacterial species, directly into the production vat (direct inoculation). In all cases, inoculation is carried out in milk that has been previously heated to fermentation temperature, which depends on the lactic acid bacteria used; this temperature is between 40 and 45°C. Depending on the type of yoghurt produced, firm or stirred, fermentation takes place in pots or in fermentation tanks.

➢ **Firm yoghurt**

Coagulation takes place directly in the jar, after aseptic packaging. Inoculation and, if necessary, the addition of flavourings or colourings are carried out in the milk. If necessary, fruit preparations are added directly to the jar. The inoculated milk is then distributed into the jars, which are then quickly sealed, incubated and cooled. Incubation takes place in hot chambers, with forced circulation of hot air, followed by cold air for cooling.

➢ **Mixed yoghurt**

Fermentation takes place in the production vat. At the end of fermentation, the gel may be stirred in the vat, then pumped out to be cooled and packaged. The vat can be multifunctional (heat treatment of the milk, fermentation, cooling).

- **Controlling and stopping fermentation**

To produce yoghurt, two conditions must be met:

- The fermentation temperature of the yoghurt is maintained between 40 and 45°C, which corresponds to the optimum temperature for the mixed culture of *S. thermophilus* and *L. bulgaricus*.
- The incubation time is 3 hours. This ensures that the culture is stopped at a given pH and avoids over-acidification.

For the production of firm yoghurts, only the temperature of the incubation chamber is controlled. The incubation time is set in advance.

For the production of stirred yoghurt, the fermentation tanks are equipped with a temperature sensor and, if necessary, a pH electrode.

f. Post-fermentation treatments

This treatment is only used in the production of stirred yoghurts.

- The coagulum is stirred before cooling. This is done either by slow stirring in the fermentation tank, or by pumping the gel. In order to smooth the gel and avoid the presence of grains in the product, the coagulum can be passed through a filter.

- The various ingredients (fruit preparations) are added after fermentation for stirred yoghurts.

g. Cooling

The cooling process must allow the temperature to drop rapidly from 40-45°C to 4°C, in order to block metabolic and enzymatic activities as quickly as possible.

- **Firm yoghurts** are cooled either in cold rooms, by circulating cooled air, or in cooling tunnels. The jars are then kept at a low temperature during storage, transport and distribution.

- **Stirred yoghurts** are cooled in two stages. The first cooling phase brings the fermented milk down from 40-45°C to around 20°C. The product is packaged at this temperature and then undergoes a second cooling phase in a pot, which lowers its temperature to 4°C. This method limits structural damage to the gel.

h. Packaging

Yoghurts are generally packaged in two types of packaging material: glass or plastic (polystyrene):

- In this case, they are decontaminated before filling by exposure to UV light or by spraying with hydrogen peroxide.

- The packaging is formed directly on the packaging machine (plastic pots). When the jars are formed on site, the heating of the plastic required to form them is enough to decontaminate them.

The jars are filled and dosed using positive displacement pumps. The jars are hermetically sealed using lids decontaminated by infrared radiation. The jars are then printed with a use-by date.

Figure 24 summarises all the stages involved in making yoghurt (firm and stirred).

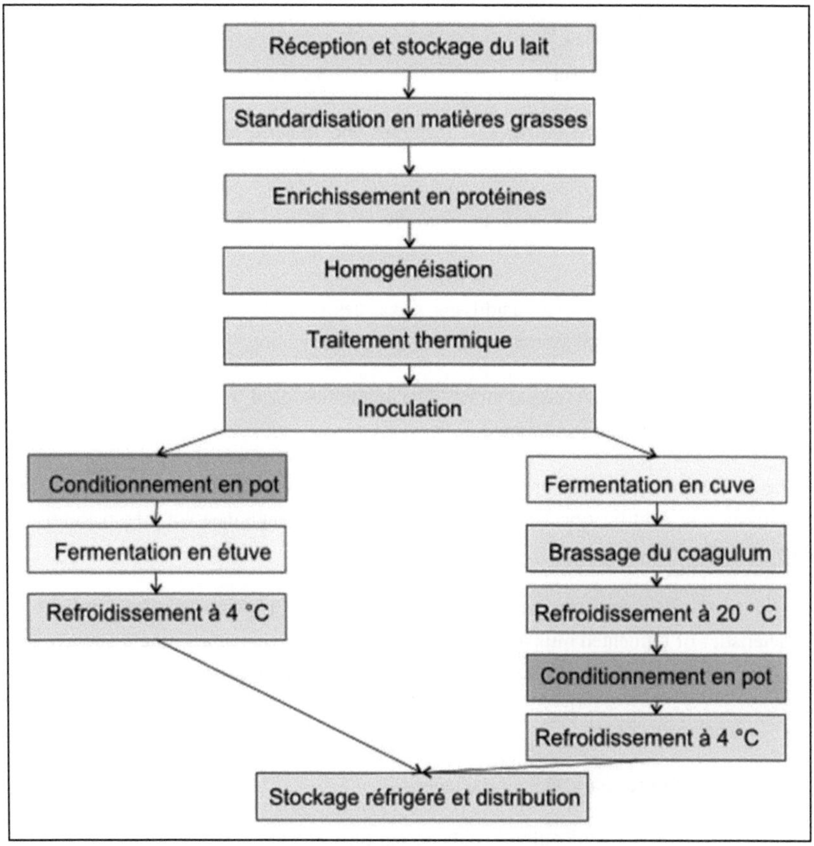

Figure 24: Diagram of yoghurt production (https://commons.wikimedia.org/wiki/File:Fabrication_yaourt.svg).

3.1.2.2. Quality control

a. Production quality control

Quality controls are carried out on raw materials and finished products, as well as during production (self-checks).

▪ Quality control of the milk used as a raw material follows the same criteria as for milk powder.

▪ Ferment quality control: The acidifying properties of each batch of lactic ferments are checked using an automated system to monitor their acidification kinetics.

▪ Quality control of fruit preparations: The microbiological risks associated with fruit preparations concern yeasts and moulds.

▪ Checks carried out during production: During production, certain critical points such as pasteurisation or acidification kinetics must be checked. Compliance with the pasteurisation schedule is checked.

▪ Inspection of finished products: The microbiological, physico-chemical and sensory characteristics of fermented milk are checked at the end of the manufacturing process (criteria according to legislation). Yoghurt must meet the following characteristics:

- ✓ Clean, even colour
- ✓ Clean taste and characteristic fragrance
- ✓ Homogeneous texture (stirred yoghurt) and firm texture (steamed yoghurt).

b. Points of sale

It's a compliance inspection:

- Exposure temperature (below 10°C) ;
- The characteristics of the packaging and the sell-by date.

c. Defects and spoilage of yoghurt

Changes in taste, appearance and texture often occur, some of which are detrimental to the final quality of the product. Among these alterations we have those reported in the following table:

Table 10: Frequent changes in yoghurt (Luquet, 1985).

	Nature	Causes
TASTE	Bitter	Too long a shelf life Proteolytic activity of ferments too high Contamination by proteolytic germs
	Yeasty, fruity, alcoholic taste	Mould contamination Poor quality fruit
	Flat taste, lack of aroma	Poor sourdough activity (short incubation period, low temperature); Dry matter content too low
	Lack of acidity	Poor yeast activity (seeding rate too low, short incubation period, low temperature)
	Too much acidity	Fermentation not carried out correctly (too high a seeding rate, high temperature); Cooling not done slowly enough
	Rancidity	Contamination by lipolytic germs
	Floury powder taste	Over-powdering
	Cooked taste	Heat treatment too high
TEXTURE	Lack of firmness (steamed yoghurt)	Inoculation too low, poor incubation (time or temperature too low), shaking before complete coagulation, dry matter too low.
	Too liquid (stirred yoghurt)	Overly violent stirring, poor incubation (low time and temperature), low dry matter, poor ferments
	Too tight	Bad ferments, T° incubation time too short
	Sandy texture	Too much heating of the milk, homogenisation at too high a temperature, too much powdering, poor stirring.
	Granular texture	Poor brewing, fat content too high

APPEARANCE	Decanting, syneresis	Over-acidification or post-acidification; Temperature too high during storage; Cooling too low; Excessive agitation and air intake; Incorrect addition of fruit or fruit pulp. Shaking yoghurts; Dry matter content too low
	Gas production	Yeast and coliform contamination
	Colonies on the surface	Contamination by yeasts and moulds
	Cream coating	Poor or no homogenisation
	Product on the lid	Poor handling
	Non-homogenised product	Bad agitation

3.2 CHEESE

According to the Codex, cheese is a ripened or unripened product with a soft, semi-hard, hard or extra-hard consistency that can be coated and in which the whey protein/casein ratio does not exceed that of milk.

Classifications

- **Depending on refining**

o Fresh or unripened" cheese is cheese that is ready for consumption shortly after manufacture.

o Ripened" cheese is cheese that is not ready for consumption immediately after production, but which must be kept for a certain period of time at the temperature and in the conditions necessary for the biochemical and physical changes characteristic of cheese to take place.

- **Depending on the consistency of the dough**

o Fresh cheese: is a cheese that is ready for consumption shortly after production without maturing.

o Soft cheese: applies to all cheeses that have not been pressed or heated during the production process:

- Bloomy rind: These are cheeses with a white, downy rind.
- Washed rind: the surface of the cheese is regularly washed with water of varying salinity and brushed to activate fermentation.

o Blue-veined cheese: the cheese is inoculated with a fungus called penicillium. This category is sometimes classed as a soft cheese.

o Pressed cheeses: the curds are pressed during the moulding process:

- Cooked paste: the curd is heated
- Uncooked cheese: the curd has not been heated.

3.2.1. Manufacturing principles

Cheese is the product obtained by coagulating the milk followed by draining the coagulum. It is essentially made up of a casein gel which retains the fat globules and a greater or lesser proportion of the aqueous phase of the milk.

There are three stages in the cheese-making process:

- coagulation or formation of the gel or coagulum;
- draining or dehydration of the gel resulting in a curd;
- maturing or enzymatic digestion of the curd.

This last stage does not exist in the case of "fromage frais" consumed after draining.

a. Milk preparation

Today, very few milks are used in their original form. Only certain cheeses have this characteristic. Industrial production requires the milk to be collected, stored and adjusted into its various components.

- **Changes in milk flora**

The milk **is** cooled in refrigerated tanks as soon as it is milked. The aim of this operation is to preserve the quality of the milk and slow down the proliferation of the bacterial species present, essentially the mesophilic lactic flora, responsible for the acidification of the milk, and the pathogenic or harmful flora.

- **Milk sanitation**

In order to sanitise the milk, heat treatment is often used, with varying degrees of intensity depending on the type of cheese: 63 to 66°C for 30 seconds for pressed cooked cheeses, 95°C for 4 to 5 minutes for fresh cheeses. This technique is carried out using pasteurisers. The choice of time/temperature combination in the cheese dairy depends on the type of product. Heating can be moderate and does not alter the cheesemaking qualities of the milk, but hygiene safety may be inadequate. These less severe heat treatments are often referred to as thermisation.

- **Standardisation of fat**

The fat content is adjusted either by adding skimmed milk to whole milk, or by adding cream to whole milk.

- **Standardisation of** protein **matter**

Protein enrichment of milk used in cheese production is designed to increase yield and consistency of cheese quality, and to improve productivity. It is achieved by adding milk powder, caseinate or whey protein to the milk. The protein content finally achieved varies according to the quality criteria sought. In most cases, they are between 37 and 45g/kg.

- **Reduced lactose content**

To limit the production of lactic acid in certain cheeses. By using membrane techniques and ultrafiltration on the milk before it goes into production.

- **Calcium rebalancing**

When milk is pasteurised, some of the soluble calcium precipitates out. To restore normal behaviour to pasteurised milk (and to cooled milk) during coagulation and draining, it is generally sufficient to add anhydrous calcium chloride ($CaCl_2$) at a maximum dose of 0.2 g/litre. Too much calcium can lead to taste defects.

b. Coagulation

Coagulation can be achieved in two ways:

- Either acid by fermentation using lactic bacteria

- Or enzymatically, using enzymes.

In traditional cheese-making techniques, the two coagulation methods are not used separately; coagulation is in fact mixed. Only the relative importance of acid coagulation and enzymatic coagulation varies. The three types of coagulum obtained :

- **Coagulums with a dominant lactic character:** lactic coagulation occurs when acidity has developed intensely (pH between 5.5 and 4.6). The dose of rennet is low and the temperature relatively low (18 to 28°C). The coagulum is crumbly, demineralised, porous and the proteins highly hydrated. The cheeses produced are mainly fresh cheeses (fromage blanc, petits suisses, etc.).

- **Coagulums with a dominant rennet character:** rennet coagulation is applied when the acidity has remained practically at the same level as that of the milk. The dose of rennet is high and the temperature close to 33°C or even higher. The coagulum obtained is elastic, supple and highly mineralised. Mechanical means are needed to remove the whey. The cheeses produced are pressed cheeses (Edam, Cantal, Cheddar, Gruyère).

- **Mixed coagulants:** mixed coagulation is obtained when the milk is moderately acidic at the time of coagulation (pH 6.5 to 5.5) and an intermediate dose of rennet is used. This range of solutions is one of the origins of cheese diversity. The cheeses produced are soft cheeses (Camembert, Brie, Munster, Bleu, etc.).

c. Draining

- **Syneresis**

Once the curds have set in the gel, they drain. This is called syneresis and corresponds to the active expulsion of part of the whey by contraction of the gel, with fine droplets spontaneously appearing on its surface. These droplets grow and join together to form a liquid film and a passive process resulting from the permeability of the coagulum, during which another part of the whey escapes. The method of coagulation has a strong influence on the structure of the gel. Gels obtained by acid coagulation have micelles associated in clusters and drain spontaneously but slowly and incompletely. On the other hand, in enzymatically produced milk, the micelles are associated in a network and form vacuoles that retain the whey; the ability of these gels to drain spontaneously is therefore poor. It is then necessary to apply physical actions such as slicing, stirring and pressing, or appropriate physico-chemical treatments such as heating.

- **Cutting**

Exudation takes place at the curd/serum interfaces; by creating more exchange by cutting the curd, draining is encouraged. Cutting can be coarse (with a ladle), regular (in 1 to 2 cm cubes) or fine (small grains). In some cases, the mixture of whey and curds is stirred to prevent the curds from sticking together and to increase the draining process.

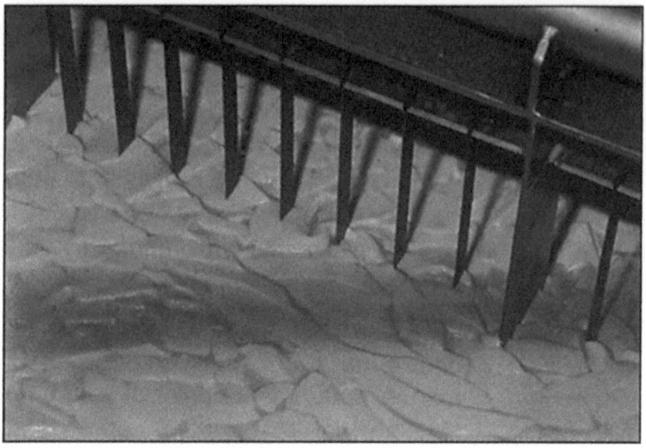

Figure 25: Cutting the curd (http://www.ulb.ac.be/sciences/cudec/LaitDerives.html).

- **Pressing**

When the aim is to achieve a low water content in the cheese, the curd is pressed into the mould after most of the whey has been removed; draining under the press generally represents only 2 to 5% of the total whey.

Figure 26: Draining with pressing
(http://www.jaimelebeaufort.fr/modeles/blog/images/fabrication_presses_pneumatiques.jpg)

- **Turnaround**

It removes whey that has accumulated in cavities, making the water content more uniform in different areas and speeding up draining (soft cheese).

Figure 27: Draining by overturning

(https://fromagedechevre.weebly.com/i-initiation-agrave-la-fabrication.html)

- **Heat treatment**

Heat promotes reactions in the coagulum, which accentuates the ability of the protein framework to contract and promote the expulsion of serum. Heating is generally used (up to a maximum of 55°C) when a high dry matter content is required in the cheese (pressed cooked cheese).

d. Salting

Cheeses are generally salted either by sprinkling dry salt on the surface, in two stages or in one, or by soaking in a brine that is often saturated for a period varying from 10 minutes to 48 hours, depending on the size of the cheese and the level of salt required. In some cases, the salt is introduced directly into the cheese mass.

Sodium chloride has several roles in cheese. It adds a salty taste and has the ability to enhance or mask the taste of certain substances formed during ripening. It modifies the hydration of the proteins, which has the effect of encouraging the drainage of the whey, resulting in additional draining and the formation of a rind on the surface of the cheese. Sodium chloride therefore has an effect on the water activity of the cheese, which determines the microbiological development both inside the paste and on the surface, as well as the chemical and biochemical reactions. As a result, the ripening of the cheese is prepared.

e. Refining

Except where the coagulum is consumed fresh, the cheese undergoes a maturing process that changes its composition, nutritional value, digestibility and organoleptic characteristics. The most obvious processes that take place are :

- Formation of a more or less hard rind which, depending on the cheese concerned, will be dry or covered with a layer of ferments or mould.

- A soft, homogenous paste forms; its colour can range from white to yellow.

- Formation of holes, cracks or fissures.

These changes are brought about by various enzymes, which come from three sources: those naturally present in the milk, added coagulating agents and those produced by various bacterial micro-organisms, yeasts and moulds.

Ripening is therefore a highly complex process due to the nature of the substrate, the diversity of the agents responsible, the variety of transformations and the number of products formed. Ripening gradually transforms the constituents of the cheese into a multitude of compounds that give the cheese its characteristics, texture, flavour and aroma. It is dominated by a number of biochemical phenomena, the most important of which are the fermentation of lactose, the enzymatic breakdown of proteins and the hydrolysis of fat.

- Lactose is hydrolysed by a number of micro-organisms. The main metabolite obtained is lactic acid in the form of lactate. Other ways of transforming lactose lead to the formation of various compounds such as CO2, water, ethanol, acetic, formic and succinic acids, alcohols and galactose. Lactate is also converted into acetic and propionic acids and CO2.

- Fat is essentially made up of triglycerides. Moulds are the most lipolytic micro-organisms in cheese. The fatty acids formed can then be modified by micro-organisms and enzymes. The compounds most often identified in cheeses are esters, methyl ketones, thioesters, secondary alcohols and lactones.

- Proteins undergo a series of transformations that result in the appearance of peptide fractions of increasingly lower molecular weight, free amino acids and compounds resulting from the catabolism of the latter. These transformations contribute to the texture and flavour of the cheese.

The main transformations of milk components during cheese preparation are shown in Figure 28.

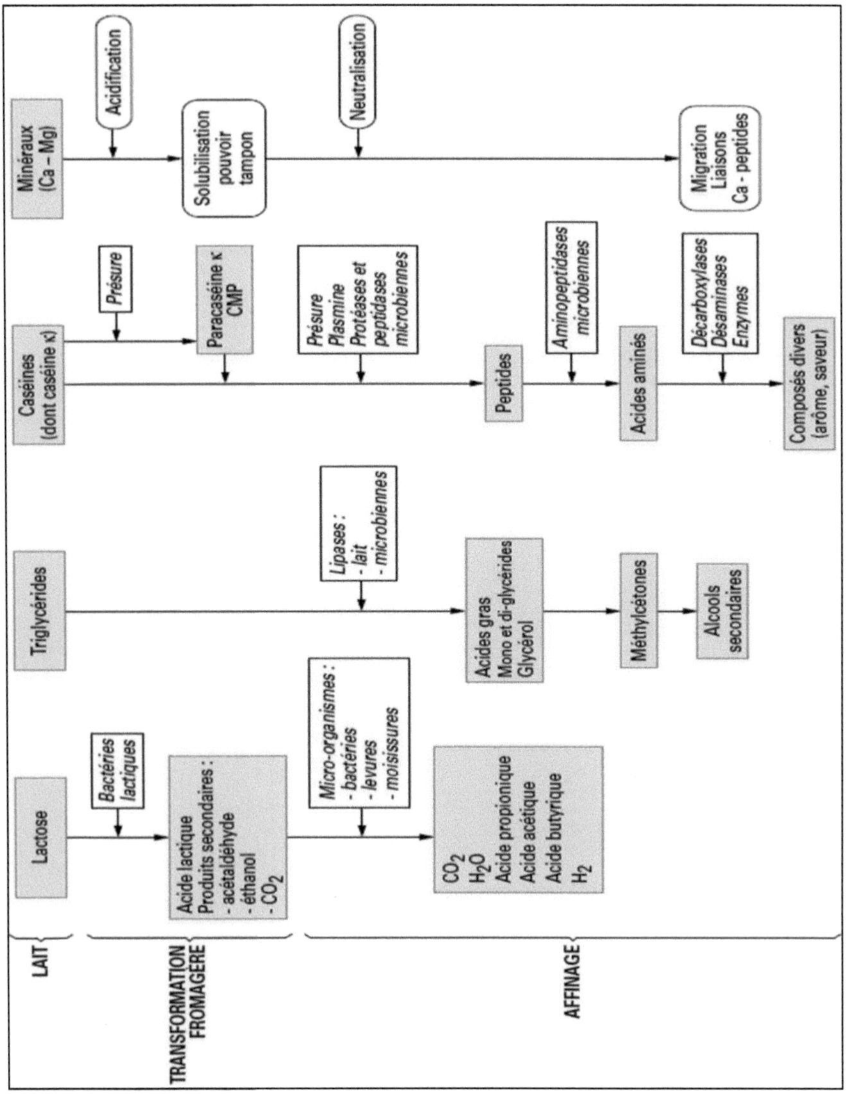

Figure 28: Main transformations of milk components during cheese production (Goudédranche et al., 2001)

Comparison of the different cheese-making stages

There is a huge variety of cheeses, and although they are classified according to the type of paste, they differ in the way they are sown, drained and matured.

Table 11 gives a summary of the different cheeses according to the type of paste, the maturing method and some of the special features of their production.

Table 11: Presentation of the different types of cheese and some of the special features of their manufacture

Type de fromage		Exemples	Coagulation	Egouttage	Salage	Affinage/(taux d'humidité)
Fromages à pâte fraîche		Petit suisse, fromages blancs	Lactique sans ou avec légère action de présure ou présure seule	Spontané 6 à 48 h, ou centrifugation	Non sauf demi-sel	Aucun (70 à 75 %)
Fromages à pâte molle	croûte fleurie	Camembert, Brie (vache), Chabichou, Sainte Maure (chèvre)	Mixte à dominante lactique	Spontané et tranchage éventuel 1 à 3 jours	Sel sec ou saumure	3 à 8 semaines 10 à 14°C (50 à 55 %)
	croûte lavée	Munster, Epoisse, Maroille, Livarot				
	pâte persillée	Roquefort (brebis), Bleu d'Auvergne (vache)	Mixte à dominante présure	Découpage et brassage 2 à 5 jours	Sel sec ou saumure	1,5 à 3 mois 6 à 10°C (50 à 55 %)
Fromages à pâte pressée	non cuite	Reblochon, Cantal	Présure	Découpage, brassage et pressage 3 à 24 h	Sel sec	3 semaines à 3 mois 10 à 14°C (45 à 50 %)
	cuite	Beaufort, Comté		Découpage, brassage, chauffage (55°C) puis pressage 24 h	Sel sec ou saumure	1 à 3 semaines 10 à 12°C puis 2 à 6 mois à 16-20°C (35 à 40 %)

3.2.2. Quality

3.2.2.1. Manufacturing defects

There is a huge variety of cheeses, and for each type of cheese there can be manufacturing defects. Here we look at some of the faults and accidents that can occur in the manufacture of soft cheeses. There are :

- Surface accidents
- Accidents in the dough

Table 12: Some defects in soft cheeses (Goudédranche et al., 2001; Desfleurs et al., 1985; Tormo et al., 2004)

Defects	Comments	Causes/Root causes	
APPEARANCE DEFECTS			
Blue	Appearance of bluish or greenish spots on the surface	Development of Penicillium expansum (glaucum) and/or contamination by Penicillium roqueforti. Contamination of the production room environment	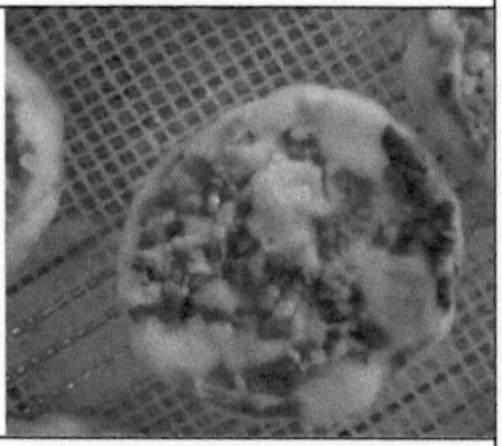
	Development of blue inside pasta Camembert-type soft cheeses	Contamination of milk by feeding cows mouldy maize silage Contamination of the atmosphere in production rooms	
Cat hair mucor	Appearance of blackish patches above the white felting	Development of mucor (Mucor mucedo). Insufficient acidification, draining and/or salting	
Greasy rind Grease or powdery mildew or "toad skin	Thick, wrinkled rind that is greasy to the touch and tends to peel off easily Penicillium camemberti has difficulty establishing itself.	Excessive development of Geotrichum candidum (Oïdium lactis). Temperature too high at the end of draining or in the hâloir.	

Black or brown spots		Contamination by Penicillium bruno-violaceum, which produces colourless puberulic acid. In the presence of iron salts, the acid turns purple-brown.	
Yellowing and browning of the crust	Rapid ageing of Penicillium camemberti, which first turns yellow and then brown	Inadequate development of *Penicillium camemberti*. Condensation under paper	

TEXTURE DEFECTS			
Dry paste	Plastery cheese, the core of the cheese remains whitish, firm and acidic	Insufficient ripening due to excessive draining Temperature in draining room too high	
Pourable paste	Cheese that "spreads" Excess moisture cannot be absorbed by moulds	Excessive development of proteolytic flora. Insufficient draining and/or too slow acidification. Presence of antibiotic or chemical residues.	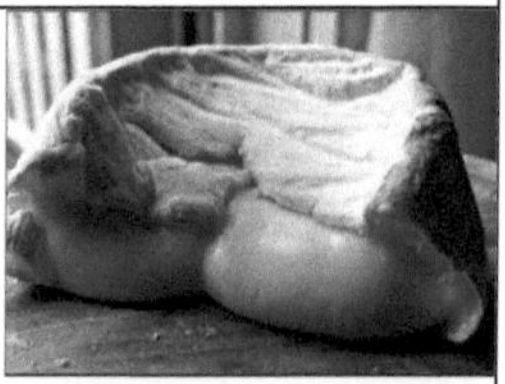
Early swelling	Numerous round holes of varying size (from the size of a pinhead to the shape of a sponge)	Coliform contamination and/or excess of heterofermentative bacteria. Delayed acidification and/or insufficient acidification. Poor draining	

TASTE DEFECTS			
Rance	Excessive lipolysis with a lack of rancid and/or soapy taste	Degradation of MG by lipases from psychotropic bacteria present in milk. Poor milking hygiene Milk storage temperature too high and/or shelf life too long. P. camemberti or roqueforti strains too lipolytic	
Bitter		Accumulation of bitter peptides in the medium. Numerous causes:	

		presence of psychotropic germs, acidification and draining too quickly, coagulants, excess calcium chloride, etc. Excessive activity of P. camemberti acid protease	
Taste of celluloid		Production of styrene by Penicillium camemberti from amino acids such as phenylalanine when the medium lacks a carbon substrate (lactic acid).	
Earthy or potato taste	Appearance of the defect in washed-rind cheeses	Contamination of milk by Pseudomonas graveolens	
Mushroom flavour		Technological origins not well known, oxidation of unsaturated fatty acids by Penicillium camemberti	

3.2.2.2. Microbiological quality

Impeccable quality milk and careful management of the microbiological process make it possible to produce quality cheese that keeps well. However, the bacterial flora present in cheeses can also originate from the production equipment and environment or come from the addition of lactic ferments during manufacture. Microbiological analyses of cheeses generally focus on germs or groups of germs that can have an impact on quality, as well as pathogens that pose a risk to consumer health. Microbiological criteria differ depending on the type of cheese.

BIBLIOGRAPHICAL REFERENCES

1. Abdullah M.A. and Alkhatib B., 2021. Some physicochemical properties of traditional sheep ghee (Samen Baladi) of pilot samples in Jordan, Review. Journal of the Saudi Society of Agricultural Sciences, 20, 111-115.

2. Abebe B., Mohammed YK., Zelalem Y., 2014. Handling Processing and utilization of milk and milk products in Ethiopia, Review. World Dairy Food, 9(2), 105-112.

3. Aljaafreh A., Al-Tahiri R., Abadleh A. and Mansour A.M., 2019. Automation of traditional butter and ghee production, Review. 45 (3), 45-52.

4. Amiot J., Fournier S., Lebeuf Y., Paquin P. and Simpson R., 2002. In: Vignola C.L. Sciences et technologiques du lait : transformation du lait, Ecole polytechniques de Québec, 600p.

5. Anonymous, 2019. Diversifying your production: cream and butter, fact sheet. https://mrepaca.fr/wp-content/uploads/2019/06/Fabriquer-de-la-creme-et-du-beurre.pdf

6. Bauman, D.E., 2000. Regulation of Nutrient Partitioning during Lactation: Homeostasis and Homeorhesis Revisited. In: Cronje, P.J., Ed, Ruman Physiology. Digestion. Metabolism and Growth and Growth and Reproduction, CAB Publishing, New York, 311-327.

7. Béal C., Sodini I., 2003. Manufacture of yoghurts and fermented milks. Techniques de l'Ingénieur, traité Agroalimentaire. Doc. F 6 315

8. Bellakhdar J., 2008. Hommes et plantes au Maghreb : éléments pour une méthode en Ethnobotanique, Maroc, Plurimondes, 386 p.

9. Benerroum N., Tamine A.Y., 2004. Technology transfer of some Moroccan traditional dairy product (lben, jben and smen) to small industrial scale. Food Microbiology, 21(4), 399-413.

10. Benkerroum N., 2013. Traditional Fermented Foods of North African Countries: Technology and Food Safety Challenges With Regard to Microbiological Risks. Comprehensive Reviews. Food Science and Food Safety, 12, 54.

11. Benkerroum N., Tamime A.Y., 2004. Technology transfer of some Moroccan traditional dairy products (Lben, Jben and Smen) to small industrial scale. Food Microbioly, 21(4), 399-413.

12. Bernard L., Leroux C. and Chilliard Y., 2008. Expression and nutritional regulation of lipogenic genes in the ruminant lactating mammary gland. Bioactive Components of Milk, In Advances in Experimental Medicine and Biology, 606, 67-108.

13. Bingham E.W. and Groves M.L., 1979. Properties of casein kinase from lactating bovine mammary gland. J. Biol. Chem. 254, 4510-4515.

14. Boisgard R. Chanat E., Lavialle F., Pauloin A., and Ollivier-Bousquet M., 2001. Roads taken by milk proteins in mammary epithelial cells. Livestock Production Science, 70, 49-61.

15. Boubekri C., Tantaoui Elaraki A., Berrada M., Benkerroum N., 1984. Caractérisation physicochimique du l'ben marocain. Le Lait, INRA Editions, 436-447.

16. Bouseta A., Selli S., Amanpour A. and TsouliSarhir S., 2021. Original Research Article, Fingerprint of aroma-active compounds and odor activity values in a traditional Moroccan fermented butter "Smen" using GC-MS-Olfactometry, 96, 1-10.

17. Boussekine R., Merabti R., Barkat M., Becila F., Belhoula N., Mounier J., Bekhouche F., 2020. Traditional fermented butter smen/dhan: Current knowledge, production and consumption in Algeria. Journal of Food Research, 9(4), 71-82.

18. Boutinaud M. and Jammes H., 2002. Potential uses of milk epithelial cells: a review. Reprod. Nutr. Dev. 42,133-147.

19. Boutinaud M., Guinard-Flament J., Jammes H., 2004. The number and activity of mammary epithelial cells, determining factors for milk production. Reprod. Nutr. Dev. 44, 499-508.

20. Boutonnier J.L., 2007. Standard cream and butter milk fat, Technique de l'ingénieur, traité Agroalimentaire. Doc F 321, 16 p.

21. Bova F., 2012. Study on the evolution and perspectives of the uses of dairy fats and proteins by the agri-food industries in the European Union. Study carried out by GEM and AND International for FranceAgriMer. Final report 152p.

22. Brenaut P., Bangera R., Bevilacqua C., Rebours E., Cebo C., Martin P., 2012. Validation of RNA isolated from milk fat globules to profile mammary epithelial cell expression during lactation and transcriptional response to a bacterial infection. Journal of Dairy Science 95, 6130-6144.

23. Brodin D., 1989. The spreadability of butter: assessment methods, improvement by fractional crystallisation. Thesis. University of Caen, France.

24. Brown D.A., 2001. Lipid droplets: proteins floating on a pool of fat. Curr. Biol. 11(11), 446-449.

25. Burgoyne R.D., and Duncan J.S., 1998. Secretion of Milk Proteins. J Mammary Gland Biol Neoplasia, 3, 275-286.

26. Bylund G., 1995. Butter and dairy spreads. Dairy processing handbook. Tetra Pak Processing Systems AB S-221 86 Lund, Sweden, 442p.

27. Capuco A.V., Wood D.L., Baldwin R., McLeod K., Paape M.J., 2001. Mammary cell number, proliferation, and apoptosis during bovine lactation: relation to milk production and effect of bST. Journal of Dairy Science, 84, 2177-2187.

28. Cayot P., and Lorient D., 1998. Structures and technofunctions of milk proteins, TEC & DOC. ed. Lavoisier, 384p.

29. Cesbron-Lavau E., Lubrano-Lavadera A.S., Braesco V., Deschamps E., 2017. Protein-rich fromage frais, petits-suisses and fermented milks. Cahiers de nutrition et de diététique 52, 33-40.

30. Chilliard Y. 2006. Ruminant feed and the fatty acid composition of their milk: what is the plasticity of the different trans isomers of C18:1 and C18:2? Sci. Aliments 26:475-479.

31. Codex Alimentarius, 1995. Codex STAN 193-1995, Codex General Standard for Butter, p 01.

32. Codex, 2011. Codex standard on fermented milks. Milk and milk products (2nd edition). www.fao.org/input/download/standards/332/CXS_206f.pdf.

33. Congrès international de la répression des fraudes, 1909. 1er Congrès pour la répression des fraudes alimentaires et pharmaceutiques, Paris.

34. Corredig M. and Dalgleish D.G., 1998. Effect of Heating of Cream on the Properties of Milk Fat Globule Membrane Isolates. Journal of Agricultural and Food Chemistry, 46(7), 2533-2540.

35. Courtin P., Monnet M. et Rul F., 2002. Cell-wall proteinases PrtS and Prt B have a different role in Streptococcus thermophiles / Lactobacillus bulgaricue mixed cultures in milk. Microbilogy, 148, 3413-3421.

36. Couture Y. and Mulon P.Y., 2005. Procedures and surgeries of the teat. Veterinary Clinics of North America: Food Animal Practice, 21(1), 173-204.

37. Couvreur S., and Hurtaud C., 2007. The milk fat globule: secretion, composition, functions and factors of variation. INRA Prod. Anim. 20(5), 369-382.

38. Croguennec T., Jeantet R., Brulé G., 2008. Fondements physicochimiques de la technologie laitière, TEC & DOC, Ed. Lavoisier, 184p.

39. Dalgleish D.G., and Corredig M., 2012. The Structure of the Casein Micelle of Milk and Its Changes During Processing. Annual Review of Food Science and Technology, 3, 449-467.

40. Danthine S., Blecker C., Paquot M., Innocente N., Deroanne C., 2000. Evolution des connaissances sur la membrane du globule gras du lait : synthèse bibliographique. Lait, 80, 209-222.

41. De Ofeu Aguiar Prado L., 2018. Prediction of milk fat production and composition by modelling: role of absorbed nutrient fluxes. PhD thesis, University of Paris-Saclay, 256p.

42. Delouis C. and Richard P., 1991. La lactation. In: Thibault C. and Levasseur M., La reproduction chez les mammifères et chez l'homme. Editors. INRA, Ellipse, Paris, France 486-514.

43. Demeffe L. and Vercaigne J P., 2016. Approche de défauts Accidents de fabrication en fromagerie. Pôle Technologique DiversiFerm. http://diversiferm.be/wp-content/uploads/2015/11/Approche-de-d%C3%A9fauts-de-fabrication-de-fromages.pdf

44. Desfleurs M., Desmazeaud M., Hardy J., Souverain R., 1985. Technological aids. In Luquet F.M., Laits et produits laitiers, Vol 2. Lavoisier TEC&DOC, Paris.

45. Dickow J.A., Larsen L.B., Hammershøj M., Wiking L., 2011. Cooling causes changes in the distribution of lipoprotein lipase and milk fat globule membrane proteins between the skim milk and cream phase. Journal of Dairy Science, 94(2), 646-656.

46. Dosogne H., Arendt J., Gabriel A., Burvenich C., 2000. Physiological aspects of milk secretion by the bovine udder, Ann. Vet. Vet, 144(6), 357-382.

47. Dudez P., Simson D., François M., 2017. Transforming fresh dairy products on the farm, Educagri, Paris, France, 128p.

48. Dylewski D.P., Dapper C.H., Valivullah H.M., Deeney J.T., Keenan T.W., 1984. Morphological and biochemical characterization of possible intracellular precursors of milk lipid globules. Eur. J. Cell Biol, 35, 99-111.

49. El-Aidie S.A-A.M., 2018, The Healthiness of Commercial Butter in Malaysia: Evaluation ofthe Physicochemical and Microbial Quality, 1(4), 1-7.

50. Evers J.M., 2004. The milk fat globule membrane compositional and structural changes post secretion by the mammary secretory cell. Int. Dairy J., 14, 661-674.

51. FAO (Food and Agriculture Organization of the United Nations), 1998. Milk and milk products in human nutrition, FAO Food and Nutrition Paper Series, No. 28. https://www.fao.org/3/t4280f/t4280f00.htm

52. FAO, 1990. The technology of traditional milk products in developing countries. Paper N◦85, Rome, Food and Agricultural Organization of the United Nations, 333p.

53. FAO, 1998. Milk and milk products in human nutrition. FAO Food and Nutrition Paper Series No. 28. Food and Agriculture Organization of the United Nations (FAO) and the Post-harvest Operations Information Network. http://www.fao.org/docrep/T4280F/T4280F00.htm.

54. FAO, 2023. Gateway to milk production and dairy products, the composition of milk https://www.fao.org/dairy-production-products/products/la-composition-du-lait/fr/

55. Federighi M., 2005. Food bacteriology compendium of food hygiene. 2, Ed.

56. Fredot E., 2005. Connaissance des aliments, bases alimentaires et nutritionnelles de la diététique, Tec & Doc, Lavoisier, Paris, 295 p.

57. Gavoye S. and Paysant B. Coagulants. History of coagulants. Revue des ENIL n° 347 / 12 - 2017.

58. Gazu L., Eshete T., Kassa G., 2018. Physicochemical analysis and microbial quality of cowbutter obtained from Menz district of Amhara region, Ethiopia. Afr. J. Bacteriol. Res. 10(3), 34-43.

59. Gille D., and Schmid A., 2015. Vitamin B12 in meat and dairy products. Nutr. Rev. 73,106-115.

60. Goudédranche H.; Camier-Caudron B.; Gassi J.Y.; Schuck P., 2001. Cheese transformation processes (parts 1 and 2). Techniques de l'Ingénieur, traité Agroalimentaire Réf : F6305 v1, F6306 v1.

61. Goudédranche H.; Camier-Caudron B., Gassi J. Y.; Schuck P., 2002. Cheese transformation processes (part 3). Techniques de l'Ingénieur, traité Agroalimentaire Ref: F6307 v1.

62. Guigue I., 2016. L'affinage : Généralités. Institut de l'élevage. http://idele.fr/reseaux-et-partenariats/reseaux-mixtes-technologiques/publication/idelesolr/recommends/laffinage-des-fromages-fermiers-lactiques.html

63. Guizani N., Haffar K., Fekih Hassen J., Karoui I., Gaida Mahjoub A. Et Zidi Y., 2012. Best available techniques (BAT) for the dairy industry. Final report, Study conducted by the Flemish Institute for Technological Research (VITO, Belgium), the Tunis International Centre for Environmental Technologies (CITET, Tunisia).

64. Haddadian Z., Bremer P., Eyres G.T., Carne A., Everett D.W., 2016. The impact of cream churning conditions on xanthine oxidase activity and oxidation-reduction potential in model emulsion systems. International Dairy Journal, 60, 55-61.

65. Harrati E., 1974. Research on Lben. Laboratoire de microbiologie, Institut National Agronomique d'Alger, 21-29.

66. Heid H.W., Keenan T.W., 2005. Intracellular origin and secretion of milk fat globules. Eur. J. Cell Biol, 84, 245-258.

67. Herve L., Quesnel H., Lollivier V., Boutinaud M., 2015. Regulation of cell number in the mammary gland by controlling the exfoliation process in milk in ruminants. J. Dairy Sci. 99, 1-10.

68. Houlihan A.V., Goddard P.A., Kitchen B.J., Masters C.J., 1992. Changes in structure of the bovine milk fat globule membrane on heating whole milk. Journal of Dairy Research 59(03), 321-329.

69. Hunziker W., and Kraehenbuhl JP., 1998. Epithelial transcytosis of immunoglobulins. J Mammary Gland Biol Neoplasia, 3, 287-302.

70. Husveth F., 2011. Physiological and reproductive aspects of animal production. Created by XMLmind XSL-FO Converter, 118 p. https://dtk.tankonyvtar.hu/xmlui/bitstream/handle/123456789/7396/0010_1A_Book_angol_05_termeleselettan.pdf?sequence=1&isAllowed=y

71. Idoui T., Benhamada N., Leghouchi E. (2010). Microbial quality, physicochemical characteristics and fatty acid composition of a traditional butter produced from cows' milk in East Algeria. Grasas y Aceites, 61(3), 232-236.

72. Jammes H., and Djiane J., 1988. The development of the mammary gland and its hormonal control in the bovine species. INRA Prod. Anim. 1, 299-310.

73. Jeantet R., Croguennec T., Mahaut M., Schuck P., Brulé G., 2008. Dairy products, Tec et Doc Ed, Lavoisier, 184p.

74. Jensen R.G. 1995. Handbook of milk composition. Academic Press, 919 p.

75. Jensen R.G. 2002. The Composition of Bovine Milk Lipids: January 1995 to December 2000. Journal of Dairy Science, 85(2), 295-350.

76. JORA (Journal officiel de la république Algérienne) 2017, Inter-ministerial order of 2 Moharram 1438 corresponding to 4 October 2016 setting the microbiological criteria for foodstuffs. Journal officiel de la république algérienne n° 39 du (8 Chaoual 1438) 2 July 2017.

77. JORA. (Official Journal of the Algerian Republic), 1998. Interministerial decree no. 96 of 10 December 1998.

78. JORF, 2007. Decree no. 2007-628 of 27 April 2007 on cheeses and cheese specialities

79. Juriaanse A.C. and Heertje I., 1988. Microstructure of shortenings, margarine and butter: a review. Food Microstructure, 7(2), 181-188.

80. Kathriarachchi K., Leus M., Everett D.W., 2014. Oxidation of aldehydes by xanthine oxidase located on the surface of emulsions and washed milk fat globules. Int. Dairy J., 37(2), 117-26.

81. Keenan T.W., Dylewski D.P., 1995. Intracellular origin of milk lipid globules and the nature and structure of the milk lipid globule membrane. In: Advanced dairy chemistry 2. Lipids, 2 Ed, Fox P.F. (Ed), Chapman and Hall, London, UK, 2, 89-130.

82. Keenan T.W., Patton S., 1995. The milk lipid globule membrane. In: Handbook of milk composition. Jensen R. G. (Ed), Academic Press, New York, USA, 5-62.

83. Keogh, M.K., 2006. Chemistry and Technology of Butter and Milk fat Spreads. Advanced Dairy Chemistry. P.F. Fox and P. L. H. McSweeney. New York, Springer. 2: lipids: 333-363.

84. Klein M.S., Almstetter M.F., Schlamberger G., Nurnberger N., Dettmer K., Oefner P.J., Meyer H.H.D., Wiedemann S., Gronwald W., 2010. Nuclear magnetic resonance and mass spectrometry-based milk metabolomics in dairy cows during early and late lactation. Journal of Dairy Science, 93, 1539-1550.

85. Kumaresh Halder, Jatindra Kumar Sahu, Satya Narayan Naik, Surajit Mandal, Subrata Kumar Bag 2021. Improvements in makkhan (traditional Indian cultured butter) production: a review J. Food Sci. Technol, 58(5), 1640-1654.

86. Lapoint-Vignola C., 2002. Dairy science and technology, milk processing. Presse internationale, Ecole polytechnique de Montréal, 600p.

87. Larpent J.P., 1996. Milk and non-fermented dairy products. In Bourgeois C.M., Mescle J.F. Zucca J., Microbiologie alimentaire tome I : Aspect microbiologique de la sécurité et de la qualité des aliments. Edit Lavoisier Tech & Doc , Paris, 671p.

88. Larsen T., 2015. Fluorometric determination of free glucose and glucose 6-phosphate in cows' milk and other opaque matrices. Food Chem, 166, 283-286.

89. Larsen T., and Fernandez C., 2017. Enzymatic-fluorometric analyses for glutamine, glutamate and free amino groups in protein-free plasma and milk. J. Dairy Res. 84, 32-35.

90. Leclercq A., 1999. Intérêt nutritionnel du lait pour l'Homme, Doctoral thesis in veterinary medicine, Faculty of Medicine, Créteil, 58p.

91. Lee S.J. and Sherbon J.W., 2002. Chemical changes in bovine milk fat globule membrane caused by heat treatment and homogenization of whole milk. Journal of Dairy Research 69(04), 555-567.

92. Leonil J., Michalski M.C., Martin P., 2013. Supramolecular structures of milk: structure and nutritional impact of casein micelle and fat globule. INRA Prod. Anim. 26, 129-144.

93. Lin, Y., Sun X., Hou X., Qu B., Gao X., Li Q., 2016. Effects of glucose on lactose synthesis in mammary epithelial cells from dairy cow. BMC Vet. Res. 12, 81.

94. Linzell J.L. and Peaker M.. 1971. Mechanism of milk secretion. Physiol Rev. 51, 564-597.

95. Linzell J.L., Mepham T.B., Peaker M., 1976. The secretion of citrate into milk. J.of Physiology, 260, 739-750.

96. Lopez C., 2011. Milk fat globules enveloped by their biological membrane: Unique colloidal assemblies with a specific composition and structure. Current Opinion in Colloid & Interface Science, 16, 391-404.

97. Lopez C., Bourgaux C., Lesieur P., Ollivon M., 2002. Crystalline structures formed in cream and anhydrous milk fat at 4°C. Lait, 82, 317-335.

98. Lopez C., Briard-Bion V., Ménard O., Beaucher E., Rousseau F., Fauquant J., Leconte N., Benoit R., 2011. Fat globules selected from whole milk according to their size: Different compositions and structure of the biomembrane, revealing sphingomyelin-rich domains. Food Chemistry, 125(2), 355-368.

99. Lopez C., Madec M.N., Jimenez-Flores R., 2010. Lipid rafts in the bovine milk fat globule membrane revealed by the lateral segregation of phospholipids and heterogeneous distribution of glycoproteins. Food Chem, 120, 22-33.

100. Lopez-Calleja I., Alonso I.G., Fajardo V., Rodriguez M.A., Hernandez P.E., Garcia T., Martin R., 2005. PCR detection of cows' milk in water buffalo milk and mozzarella cheese. International Dairy Journal, 15, 1122-1129.

101. Luquet F.M., 1985. Lait et produits laitiers : vache-brebis-chèvre, vol 1 : Les laits de la mamelle à la laiterie. Lavoisier Tec & Doc, Paris, 397p.

102. Mahaut M., Jeantet R., Bruleg G., Schuch P., 2000. Les produits industriels laitiers, Ed Tec & Doc, Lavoisier, Paris, 178p.

103. Mather I.H. and Keenan T.W., 1998. Origin and secretion of milk lipids. J.A.M.A., 3, 259-273.

104. Mathieu J., 1997. Initiation à la physicochimie du lait, Lavoisier TEC&DOC, Paris, 220p.

105. McMahon D.J. and McManus W.R., 1998. Rethinking Casein Micelle Structure Using Electron Microscopy. J. Dairy Sci. 81, 2985-2993.

106. McPherson A.V., Dash M.C., Kitchen B.J., 1984. Isolation and composition of milk fat globule membrane material: I. From pasteurized milks and creams. Journal of Dairy Research, 51(2), 279-287.

107. Mekdes A., 2008. Assessment of processing techniques and quality attributes of butter produced in delbo watershed of Wolayita zone southern Ethiopia, Thesis Hawassa University, Ethiopia, 63p.

108. Meshref A.M.S., 2010. Microbiological quality and safety of cooking butter in Beni-Suef governorate-Egypt. African Health Sciences, 10(2), 193-198.

109. Michalski M.C. and Parmentier M., 2003. Fractionation and functionalities of globular and anhydrous milk fat. Journ. Filières Lait, Rennes, France, 76-86.

110. Michalski M.C., Briard V., Michel F., 2001. Optical parameters of milk fat globule for laser light scattering measurements. Lait, 81, 787-796.

111. Morin P., Jiménez-Flores R., Pouliot Y., 2007. "Effect of processing on the composition and microstructure of buttermilk and its milk fat globule membranes." International Dairy Journal 17(10), 1179-1187.

112. Mortensen B.K., 2011. Butter and other milk fat products. The Product and Its Manufacture. Encyclopedia of Dairy Sciences (Second Edition). Academic Press, 492-499.

113. Mulder, H. and P. Walstra (1974). The milk fat globule - Emulsion science as applied to milk products and comparable foods. Editions Commonwealth Agricultural Bureau, England, and Centre of Agricultural Publishing and documentation, Netherlands, 300p.

114. Murphy M.A., Shariflou M.R., Moran C., 2002. High quality genomic DNA extraction from large milk samples. Journal of Dairy Research, 69, 645-649.

115. Neville M., 1995. The physical properties of human and bovine milks In Jensen R. Handbook of milk composition-General description of milks, Academic Press, Inc 82, 919.

116. Neyer F., 2016. Les ferments lactiques, Généralités. Revue des ENIL n° 345, 2-3.

117. Ngounou C., Ndjouenkeu R., Mbofung F. and Noubi I., 2003. Demonstration of calcium and magnesium bioavailability during milk fermentation by lactic acid bacteria isolated from Zebu curdled milk. Journal of food Engineering, 57, 301-307.

118. Ollivier-Bousquet M., 1998. Transferrin and prolactin transcytosis in the lactating mammary epithelial cell. J. Mammary Gland Biol. Neoplasia, 3, 303-313.

119. Ouadghiri M., 2009. Biodiversity of lactic acid bacteria in raw milk and its derivatives 'Lben' and 'Jben' of Moroccan origin. Université Mohammed V - Agdal, 29.

120. Pointurier H., 2003. La gestion matière dans l'industrie laitière, Tec & Doc, Lavoisier, France, 388p.

121. Pointurier H., Adda J., 1969. Butter qualities and defects. In: Beurrerie Industrielle: science et technique de la fabrication du beurre. La Maison Rustique (Ed), Paris, France, 288-330.

122. Poutrel B., Caffin J.P., Rainard P., 1983. Physiological and Pathological Factors Influencing Bovine Serum Albumin Content of Milk. J. Dairy Sci. 66, 535-541.

123. Ramet J.P., 1985. Cheese-making and cheese varieties in the Mediterranean basin. Food and Agriculture Organization of the United Nations, Rome.

124. Renner E., 1989. Micronutrients in milk and milk-based food products. Elsevier, Applied Science, London, 311p.

125. Richoux R. Coagulants. The different families of coagulants and how they are obtained. Revue des ENIL n° 347 / 12 - 2017.

126. Ronez F., 2012. Functional exploration and industrial valorisation of the bacterial heat shock protein Lo18. Doctoral thesis, École doctorale Environnements, Santé.

127. Saleem Jabir Ahmed S., Osman Mohamed Abdalla M., Abdalla Rahamtalla S., 2016. Microbiological Quality of Cows' Milk Butter Processed in Khartoum State, Sudan. British Microbiology Research Journal, 11 (1), 1-10.

128. Samet-Bali O., Ayadi M.A., Attia H., 2009. Traditional Tunisian butter: Physicochemical and microbial characteristics, 42, 899-905.

129. Sarab M.L., Nouri M., Tarighat-Esfanjani A., 2019. Characteristics of Iranian traditional butter produced in Sarab in comparison with European and national standards. Progress in Nutrition, 21, Supplement 1, 416-421.

130. Seminel L., 2016. Le livre blanc des pâtes pressées de montagne, la patience du fromager. Cheeses & Chefs www.fromagesetchefs.com

131. Shennan D.B., and Peaker M., 2000. Transport of Milk Constituents by the Mammary Gland. Physiological Reviews, 80, 925-951.

132. Shingfield K.J., Lee M.R.F., Humphries D.J., Scollan N.D., Toivonen V., Reynolds C.K. Beever D.E., 2010. Effect of incremental amounts of fish oil in the diet on ruminal lipid metabolism in growing steers. British Journal of Nutrition, 104, 56-66.

133. Smith-Kirwin S.M., O'Connor D.M., Johnston J., De Lancey E., Hassink S.G., Funanage V.L., 1998. Leptin expression in human mammary epithelial cells in breast milk. J. Clin. Endocrinol. Metab 83, 1810-1813.

134. Tantaoui-Elaraki A., Berrada M., El Marrakchi A., Berramou A., 1983. Etude sur le lben marocain. Le Lait, INRA Editions, 63, 230-245.

135. Tantaoui-Elaraki A., El Marrakchi A., 1987. Study of the Moroccan dairy products: Lben and smen. World Journal of applied Microbioly and Biotechnology, 211-220.

136. Tao N., De Peters E.J., Freeman S., German J.B., Grimm R., Lebrilla C.B., 2008. Bovine Milk Glycome. Journal of Dairy Science, 91, 3768-3778.

137. Tessier L., Technologies des bioprocédés industriels.2ème Edition, Collège Shawinigan, 2263, avenue du Collège, Shawinigan, Québec, Canada 2018, pp 492.

138. Tormo H. et al., 2004. Guide d'appui technique pour l'accident de fromagerie à la ferme "le poil de Chat ou mucor". Dairy technology [electronic document]. Lyon, Fédération des éleveurs de chèvre (FNEC), Institut de l'élevage and Centre fromager de carmejane. http://www.accidentfromagerie.fr/IMG/pdf/Guide_Mucor.pdf.

139. Valivullah H.M., Dylewski D.P., Keenan T.W., 1986. Distribution of terminal transferases of acylglycerol synthesis in cell fractions from lactating mammary gland. Int. J. Biochem, 18, 799-806.

140. Walstra P., Geurts T.J., Noomen A., Jellema A., Van Boekel M.A.J.S., 1999. Dairy Technology: Principles of milk properties and processes. Ed. Marcel Dekker, 727p.

141. Walstra P., Wouters J.T.M., Geurts T.J., 2005. Colloidal Particles of Milk. Dairy Science and Technology, Second Edition, CRC Press, 109-158.

142. Wasswa J., Sempiira E.J., Mugisa D.J., Muyanja C., Kisaalita W.S., 2017. Quality Assessment of butter produced using traditional and mechanized churning methods. Afr. J. Food Agric. Nutr. Dev, 17(1), 11757-11770.

143. Wattiaux M.A., 1997. Dairy essentials: Nutrition and feeding, reproductionand genetic selection, lactation and milking, raising dairy heifers. Ed. The Babcock Institute for international Dairy research and development, Madison Wisconsin, 69p.

144. Wendrinska A., Addey C.V.P., Orange P.R., Boddy L.M., Hendry K.A.K., Wilde C.J., 1993. Effect of a milk fat globule membrane fraction on cultured mouse mammary cells. Biochem Soc Trans, 21, 220S.

145. Yao Y., Zhao G., Yan Y., Mu H., Jin Q., Zou X., Yang X., 2016. Milk fat globules by confocal Raman microscopy: differences in human, bovine and caprine milk. Food Res Int, 80, 61-69.

[illegible]e A., Singh H., Oldfield J.D. Anema S., 2004a. Kinetics of heat-induced association of [illegible]oglobulin and α-lactalbumin with milk fat globule membrane in whole milk. [illegible]tional Dairy Journal, 14(5), 389-398.

147. Ye A., Singh H., Taylor M.W. Anema S. 2002. Characterization of protein components of natural and heat-treated milk fat globule membranes. International Dairy Journal, 12, 393-402.

148. Ye A., Singh H., Taylor M.W., Anema S., 2004b. Interactions of whey proteins with milk fat globule membrane proteins during heat treatment of whole milk. Le Lait, 84(3), 269.

149. Zahara L. and Zinewi H., 2015. The microbial properties of dairy products in Ethiopia: A review. Inter. J. of Dairy science and Technol, 2(1), 088-094.

150. Zazcek M., Keenan T.W., 1990. Morphological evidence for an endoplasmic reticulum origin of milk lipid globules using lipid-selective staining procedures. Protoplasma, 159, 179-182.

151. Zulak IM and Keenan TW., 1983. Citrate accumulation by a Golgi apparatus-rich fraction from lactating bovine mammary gland. Int J. Biochem, 15, 747-750.

MIX
Papier aus verantwortungsvollen Quellen
Paper from responsible sources
FSC® C105338

d by Books on Demand GmbH, Norderstedt / Germany